Sources of Behavioral Variance in Process Safety

Process safety management seeks to establish a multi-level system to assess, document, maintain, and inspect equipment and work practices integral in controlling highly toxic and/or reactive materials. In a highly engineered environment, any variance can set off a chain of events that increases the probability of a process safety incident as violent as an explosion. Human behavior is often the biggest source of this variance, but it can also be the biggest asset for process safety management. Process industries are looking to understand sources of behavioral variance and build better processes based on sound behavioral science. Because of this clear link between behavior and process safety performance, the behavior science community has been challenged to research the behavioral root causes leading to variation that threaten process safety; create and evaluate behavioral interventions to mitigate this variation; and identify the system factors that would influence the behaviors necessary to promote process safety. This book seeks to translate behavior analysis into practical systems that can help reduce human suffering from catastrophic process safety events.

The chapters in this book were originally published in various issues of the *Journal of Organizational Behavior Management.*

Timothy D. Ludwig is a Distinguished Graduate Professor at Appalachian State University, USA, and serves on the Cambridge Center for Behavioral Studies' Commission that accredits best-in-industry safety programs. He is the author of dozens of scholarly articles and books that empirically document methods to improve safety and quality in industry through behavior analysis.

Sources of Behavioral Variance in Process Safety

Analysis and Intervention

Edited by
Timothy D. Ludwig

Routledge
Taylor & Francis Group
LONDON AND NEW YORK

First published 2018
by Routledge
2 Park Square, Milton Park, Abingdon, Oxon, OX14 4RN, UK

and by Routledge
711 Third Avenue, New York, NY 10017, USA

Routledge is an imprint of the Taylor & Francis Group, an informa business

British Library Cataloguing in Publication Data
A catalogue record for this book is available from the British Library

ISBN 13: 978-1-138-49333-9

Typeset in Minion
by RefineCatch Limited, Bungay, Suffolk

Publisher's Note
The publisher accepts responsibility for any inconsistencies that may have arisen during the conversion of this book from journal articles to book chapters, namely the possible inclusion of journal terminology.

Contents

Commentary

Citation Information

The chapters in this book were originally published in the *Journal of Organizational Behavior Management (JOBM)*. Chapter 5 was published in a 2016 special issue which was edited by Ramona A. Houmanfar. Chapter 1 was published in a 2015 special issue which was co-edited by Ramona A. Houmanfar and Mark A. Mattaini. Chapter 7 appeared in a 2017 special section of the journal, which was co-edited by Ramona A. Houmanfar and Mark A. Mattaini. All other chapters appeared in a 2017 special issue edited by Timothy Ludwig.

Chapters 1 and 7 were also included in Houmanfar and Mattaini's co-edited book *Leadership and Cultural Change: Managing Future Well-Being* (Routledge, 2018): ISBN: 978-1-138-56061-1.
Webpage: https://www.routledge.com/products/9781138560611

The chapters in this book were originally published in various issues of the *Journal of Organizational Behavior Management*. When citing this material, please use the original page numbering for each article, as follows:

Editorial
Process Safety: Another Opportunity to Translate Behavior Analysis into Evidence-Based Practices of Grave Societal Value
Timothy D. Ludwig
Journal of Organizational Behavior Management, volume 37, issues 3–4 (September 2017), pp. 221–223

Chapter 1
An Industry's Call to Understand the Contingencies Involved in Process Safety: Normalization of Deviance
Kevin Bogard, Timothy D. Ludwig, Chris Staats, and Danielle Kretschmer
Journal of Organizational Behavior Management, volume 35, issues 1–2 (June 2015), pp. 70–80

Chapter 2
Process Safety Behavioral Systems: Behaviors Interlock in Complex Metacontingencies
Timothy D. Ludwig
Journal of Organizational Behavior Management, volume 37, issues 3–4 (September 2017), pp. 224–239

Chapter 3
Complacency in Process Safety: A Behavior Analysis Toward Prevention Strategies
Cloyd Hyten and Timothy D. Ludwig
Journal of Organizational Behavior Management, volume 37, issues 3–4 (September 2017), pp. 240–260

Chapter 4
Behavioral Perspectives on Variability in Human Behavior as Part of Process Safety
Angela R. Lebbon and Sigurdur O. Sigurdsson
Journal of Organizational Behavior Management, volume 37, issues 3–4 (September 2017), pp. 261–282

Chapter 5
A Behavioral Interpretation of Situation Awareness: Prospects for Organizational Behavior Management
Kenneth Killingsworth, Scott A. Miller, and Mark P. Alavosius
Journal of Organizational Behavior Management, volume 36, issue 4 (December 2016), pp. 301–321

Chapter 6
Assessing and Preventing Serious Incidents with Behavioral Science: Enhancing Heinrich's Triangle for the 21st Century
Terry McSween and Daniel J. Moran
Journal of Organizational Behavior Management, volume 37, issues 3–4 (September 2017), pp. 283–300

Chapter 7
Leadership and Crew Resource Management in High-Reliability Organizations: A Competency Framework for Measuring Behaviors
Mark P. Alavosius, Ramona A. Houmanfar, Steven J. Anbro, Kenneth Burleigh, and Christopher Hebein
Journal of Organizational Behavior Management, volume 37, issue 2 (June 2017), pp. 142–170

Chapter 8
Integrating Behavioral Science with Human Factors to Address Process Safety
Manuel A. Rodriguez, John Bell, Michelle Brown, and Donna Carter
Journal of Organizational Behavior Management, volume 37, issues 3–4 (September 2017), pp. 301–315

Chapter 9
Leadership's Role in Process Safety: An Understanding of Behavioral Science Among Managers and Executives Is Needed
Nicole Gravina, Bob Cummins, and John Austin
Journal of Organizational Behavior Management, volume 37, issues 3–4 (September 2017), pp. 316–331

Chapter 10

Commentary: Integrating Behavioral Science with Process Safety Management
Joseph Dagen and Marcin Nazaruk
Journal of Organizational Behavior Management, volume 37, issues 3–4 (September 2017), pp. 332–338

Chapter 11

Commentary: Is Organizational Behavior Management Enough? How Language and Person-States Could Make a Difference
E. Scott Geller
Journal of Organizational Behavior Management, volume 37, issues 3–4 (September 2017), pp. 339–346

Chapter 12

Commentary: Process Safety: Look Looking Beyond Personal Safety to Address Occupational Hazards and Risks
Oliver Wirth
Journal of Organizational Behavior Management, volume 37, issues 3–4 (September 2017), pp. 347–355

For any permission-related enquiries please visit: http://www.tandfonline.com/page/help/permissions

INTRODUCTION

Process Safety: Another Opportunity to Translate Behavior Analysis into Evidence-Based Practices of Grave Societal Value

Our society often turns to engineering when the stakes are high and human lives are in the balance. The vehicles and railroads that cross our continents, the buildings that we live and work in, and the instruments used to heal are all highly engineered products that enhance human existence. To produce these, we rely on engineering related to the extraction of raw materials from our earth and the refinement the materials construct and power our modern world. As our society exploits these engineering feats, we are also frequently made aware of how these industries can cause great harm through environmental releases, explosions, and work-related fatalities. This is the new world of Process Safety.

Industrial equipment is engineered to control and transform highly hazardous materials. To ensure the integrity of this engineering, human work processes are adopted. Process safety is the management of processes that establish a multi-level system to assess, document, maintain, and inspect equipment and work practices integral in controlling highly toxic and/or reactive materials. In a highly engineered environment, any variance can set off a chain of events that increases the probability of a process safety incident as violent as an explosion. Human behavior is often the biggest source of this variance, but it can also be the biggest asset for process safety management.

Predictably, the growing field of process safety management is dominated by engineering practices such as equipment design, operating procedures, preventive maintenance systems, and quality assurance. These engineered systems are highly dependent on human behavior. Thus, engineers and managers are looking to the behavioral sciences to help them design better systems to reduce, and in some cases actually increase, human variation within their processes.

This special issue of the *Journal of Organizational Behavior Management* (*JOBM*) was instigated by the petrochemical industry who reached out to the Cambridge Center for Behavioral Studies (CCBS) in search of behavioral solutions to their process safety challenges around constructs such as normalization of deviance and complacency. Process industries are looking to understand sources of behavioral variance and build better processes based on sound behavioral science. Because of this clear link between behavior and process safety performance, the behavior analytic community has been challenged to: a) research the behavioral root causes leading to variation that threaten process safety, b) create and evaluate behavioral interventions to mitigate this variation, and c) identify the system factors that would influence the behaviors necessary to promote process safety (Bogard, Ludwig, Staats, & Kretchmer, 2015).

The contents of this *Special Issue* are a first step in translating behavior analysis into practical systems that can help reduce human suffering from catastrophic process safety events. In other areas of public health there is much discussion of translating basic research to practice to institutionalization (Davis et al., 2003; Geller, Winett, & Everett, 1982; Green & Glasgow, 2006). At the beginning of this pathway basic research is first applied through efficacy interventions to test and adapt basic principles in the uncontrolled environments of the real world. The *Journal of Organizational Behavior Management* has been producing this type of research since its inception. What we hope to do with this special issue is to generate applications of basic behavior analysis to test the efficacy of initial interventions in impacting process safety variables. However, if we stop there we won't answer the call. We need to pursue effectiveness research where we develop packages of interventions and prove their effectiveness with evidence of not only incident prevention but also sustained adoption.

Behavior analysis has done this before as we seek to reduce human suffering in all its forms. Most related is the effective and sustained application of behavior analysis to personal safety, those actions that help workers avoid and protect themselves from hazards. The resulting behavioral systems have been referred to as *Behavioral Safety* or *Behavior-Based Safety*. However, the behaviors and contingencies involved in personal safety are different from those involved in process safety. Personal safety behaviors occur in contexts characterized by repetitive work using active response classes (e.g., putting on protective equipment, maintaining posture) generalized across work settings. Behavioral safety systems that have worked in these contexts rely on applied research methodologies of direct behavioral observation, data collection, and analyses paired with social contingencies such as individual and group feedback.

The context of Process Safety is qualitatively difference. The work context is often unique, direct-acting contingencies absent, protective behaviors passive (e.g., observing gauges, inspections), and intervention success is mitigated by extensive interlocking behavioral contingencies from many other actors in complex metacontingencies. The interventions that may make up an effective behavioral system for process safety must deal with behavioral constructs such as fluency, extinction, and stimulus control.

The collection of papers assembled begin by attempting to translate basic principles of behavior analysis to the challenges associated with process safety. Cloyd Hyten and Timothy Ludwig apply numerous behavioral processes such as habituation, extinction, unprogrammed reinforcement, and rule-governed behavior to a pattern of behavioral variability colloquially termed "complacency". In the second paper, Timothy Ludwig offers a description of different behavioral classes contributing to process safety before providing a behavioral systems analysis to highlight the metacontingencies and interlocking behavioral contingencies impacting process safety performance. Then Angela Lebbon and Sigurdur Sigurdsson add a discussion of risk discounting to the behavioral decision making involved in behavioral variation leading to risk taking.

The second series of papers attempt to set up the models within which behavior analysis' applied researcher can build effective and sustainable interventions. Terry McSween and Daniel Moran enhance a decades-old safety model (i.e., Heinrich's Safety Triangle) to account for the precursors of the serious incidents targeted by process safety and describe how they might lead to better analysis of causal system failures that drive behavior. Manuel Rodriguez, John Bell, Michelle Brown, and Donna Carter introduce the behavioral audience to the field of Human Factors where human errors are the focus of process safety as a pathway to integrate these two behavioral science methodologies. Finally, most of the papers in this special issue identify leadership behaviors as a target of intervention for successful behavior analytic interventions. Nicole Gravina, Bob Cummins, and John Austin describe a tested leadership intervention where those whose behavior create the powerful metacontingencies are taught behavior science techniques to evaluate the impact of their own behavior and modify the behavior of others that maintain process safety effectiveness.

We hope that readers of this special issue will follow up on the principles and tactics offered across the spectrum of translational research. We should endeavor to demonstrate once again the robustness of our science and reach of our application. We have another opportunity to translate behavior analysis into evidence-based practices of grave societal value.

References

Bogard, K., Ludwig, T. D., Staats, C., & Kretchmer, D. (2015). An Industry's Call to Understand the Contingencies involved in Process Safety: Normalization of Deviance and Interlocking Contingencies. *Journal of Organizational Behavior Management*, *35*, 70–80. doi:10.1080/01608061.2015.1031429

Davis, D., Evans, M., Jadad, A., Perrier, L., Rath, D., Ryan, D., ... Zwarenstein, M. (2003). The case for knowledge translation: Shortening the journey from evidence to effect. *British Medical Journal*, *327*, 33–35. doi:10.1136/bmj.327.7405.33

Geller, E. S., Winett, R. A., & Everett, P. B. (1982). *Preserving the environment: New strategies for behavior change.* New York: Pergamon Press.

Green, L., & Glasgow, R. (2006). Evaluating the relevance, generalization, and applicability of research: Issues in external validation and translation methodology. *Evaluation & the Health Profession*, *29*(1), 126–153. doi:10.1177/0163278705284445

Timothy D. Ludwig

An Industry's Call to Understand the Contingencies Involved in Process Safety: Normalization of Deviance

KEVIN BOGARD

TIMOTHY D. LUDWIG

CHRIS STAATS

DANIELLE KRETSCHMER

Marathon Petroleum Company (MPC), Illinois Refining Division (IRD) adopted a behavior science approach to its safety operations becoming one of the first sites accredited for its behavioral safety program by the Cambridge Center for Behavioral Studies (CCBS). Beyond success in managing personal safety, there is increased and intense attention toward Process Safety in the oil and gas industry where equipment, processes, and behavior are managed to reduce the potential for catastrophic loss, damage, and impact on human life and livelihood. The oil and gas industry is increasingly looking to the behavior science community to understand the contingencies related to "normalization of deviance", where behaviors begin to drift from process standards and become the norm among work teams over time. Further, the oil and gas industry seeks to understand how interlocking contingencies may both shape and maintain normalization of deviance, as well as how systemic interventions can address the issue.

Portions of this manuscript were presented at the Symposium for Leadership and Cultural Change, May 23, 2014, Chicago, Illinois.

Marathon Petroleum Company (MPC) is a Fortune 30 company and is Responsible Care Certified (American Chemistry Council, 2014) for its environmental stewardship, safety programs, process safety management systems, security initiatives, and product quality. The Illinois Refining Division (IRD) has maintained the Occupational Safety and Health Administration's (OSHA) Voluntary Protection Program (VPP) STAR Site since 1999 (OSHA, 2014a).

This safety performance was not always the standard at MPC's IRD. OSHA requires that injuries needing care beyond first aid be recorded and submitted as a ratio, in which total cases are multiplied by 200,000 and then divided by exposure hours (Bureau of Labor Statistics, 2013). In 1995, the OSHA recordable rate at IRD was 3.63 for refinery employees, which was roughly consistent with the industry norm. Investigation into the factors maintaining this rate indicated that trust and communication between management and the hourly workforce was generally viewed as low, and was seen as a barrier to reducing injuries.

An employee-driven team was formed in 1996 and implemented a behavioral safety program in 1997 named Areas Communicating Trust in Safety (ACTS). The ACTS process combined research and practice from behavioral science that built behavioral safety (Geller et al., 1990; Hermann, Ibarra, & Hopkins, 2010; Ludwig & Geller, 2000; McSween, 1995; Myers, McSween, Medina, Rost, & Alvero, 2010; Sulzer-Azaroff & Austin, 2000) with practices from Total Quality Management (Deming, 1982) and other team-based performance improvement movements that seek to empower employees. The ACTS process is not run from the top down; it is owned and managed by hourly personnel. Managers, supervisors, and foreman are not in charge of the ACTS process, they are simply a part of it.

Data on safety and at-risk behaviors are collected by trained observers performing peer-to-peer job observations. These observers provide feedback on their peer's performance and identify barriers—antecedents and consequences—associated with these actions. Data are then analyzed to determine the highest frequency of at-risk behaviors among the workforce or work group. These behaviors and their barriers are reviewed in detail during employee safety meetings and with management. Problem solving around the at-risk behaviors at these meetings leads to targeted interventions in the refinery, including focused safety meetings, training initiatives, and the purchasing of new safety equipment. As an example, based on its data, the ACTS team discovered a trend indicating that workers were engaging in more at-risk behaviors and sustaining injuries an hour before and after lunch, as well as an hour before they left for the day. To combat this problem, the ACTS team held brief safety meetings to discuss pertinent safety topics based on the data on at-risk behaviors occurring around lunch and the end of the shift. This was associated with a decrease in recordable incidents from six per year to zero during this period. Trend analyses such as this are also given to shift supervisors to present the information to their respective work groups. This cycle of collecting and distributing information creates a proactive approach to improving safety.

It is standard practice in the oil and gas industry, along with most industries, not to include contractor workforce injuries in published safety statistics. However, MPC's IRD chose to report contractor injuries in addition to its own and commit to helping its contractors with implementing and refining their own behavioral safety processes. Contractors attended a 4-hr class on the ACTS behavior-based safety process and were encouraged by the refinery general manager to stop jobs if they felt the task put themselves or their coworkers at risk. After including contractor injury statistics, training their personnel, and using their dedicated observers to collect and distribute data, they experienced an initial 30% to 40% decrease in first aids and incidents.

To formalize contractor participation in behavioral safety, MPC's IRD began to sponsor another committee in 2005 run by contractor employees to manage their own behavioral safety processes. The Contractor Advisory Panel is made up of 11 contract companies that participate in similar behavioral safety processes. A substantial increase in behavioral observations was realized subsequent to the formal onset of contractor behavioral safety processes. Since then, MPC's IRD and its contractors have recorded tens of thousands of observations annually (28,883 in 2013) as the number of safety incidents has declined (see Figure 1).

Since 2005, MPC's IRD and its contractors have succeeded in keeping their injury rate below 0.50 in five separate years. In 2013, IRD's OSHA recordable rate was 0.40 (0.00 for MPC employees and 0.84 for the contractor workforce). Because of the substantial reduction in injuries associated with their application of behavioral science to safety programming, the Cambridge Center for Behavioral Studies (CCBS) has accredited MPC's IRD as a best-in-practice behavioral safety program since 2005 (CCBS, 2014). In addition, nine of the contractors participating in the Contractor Advisory Panel have also been accredited by CCBS or have received a certificate in pursuit of accreditation.

PROCESS SAFETY

Maintaining this level of excellence is challenging, as each year brings new obstacles. However, one constant in the oil and gas industry, as in other

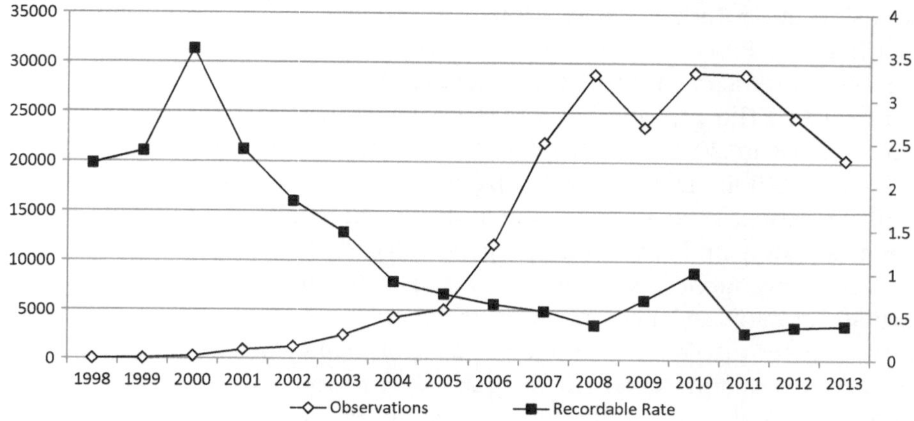

FIGURE 1 Marathon Petroleum Company, Illinois Refining Division, number of behavioral safety observations (left axis) overlaid with Occupational Safety and Health Administration recordable rates (right axis), from 1998 through 2013, for both Illinois Refining Division employees and contractor workforces combined.

industries, is the threat of catastrophic incidents. The focused management of these potential incidents is an area referred to as process safety management (OSHA, 2014b). Whereas personal safety pertains to an injury that impacts an individual, process safety incidents occur when many individuals are injured because of a problem resulting from equipment. Regarding process safety incidents, people on a worksite and in the surrounding community may become seriously injured and fatalities may occur because of breakdowns in process safety. Environmental disasters are also associated with incidents. Although equipment failure is often a cause of process safety incidents, human behavior is often associated with the event. The interaction of equipment failure and behavioral errors can result in an incident, sometimes undetectable, that can then cascade into larger and sometimes fatal consequences. Process safety impacts workers, the community, and the industry.

Process safety has changed the business landscape for many industries, especially oil and gas. Society has a negative view of oil and gas industry companies due to a negative public image resulting from process safety issues. This includes injuries, fatalities, and environmental disasters. Impacts of these events are widespread, and people have become less tolerant of these issues—as they should. Fifteen people were killed in an explosion at the BP Texas City refinery in 2005 (Mufson, 2007). Three were seriously injured in the Valero refinery propane fire in 2007 (U.S. Chemical Safety and Hazard Investigation Board, 2008), two others were seriously injured in the Veolia flammable vapor explosion in 2009 (U.S. Chemical Safety and Hazard Investigation Board, 2010), and one person was critically injured in a CITGO refinery hydrofluoric acid release in 2009 (U.S. Chemical Safety and Hazard Investigation Board, 2009). Eleven were killed in a vapor cloud explosion at an Indian Oil Corporation terminal in Jaipur in 2009 (Center for Chemical Process Safety, 2012). Seven died in an explosion and fire at a Tesoro refinery in 2010 (U.S. Chemical Safety and Hazard Investigation Board, 2014). The explosion on the Deepwater Horizon in the Gulf of Mexico cost 11 lives in 2010, catastrophic environmental damage, and billions of dollars to BP Oil (Deepwater Horizon Study Group, 2011). Recently, West Fertilizer experienced an explosion and fire that resulted in 14 fatalities (U.S. Chemical Safety and Hazard Investigation Board, 2013).

The very public investigations of these disasters have indicated that a normalization of deviance often occurs. Often several safeguards break down and are never discovered because of lack of vigilance. The root causes can be traced to actions taken in the past by previous workers. Management systems break down, and the leadership is often unaware of the hazard and growing problem. The prevention and mitigation actions that leaders think are being done are not because systems are cumbersome to workers and shortcuts have replaced the engineered process. Thus, deviant behaviors, through either intentional shortcuts or unintentional omissions, contribute to process safety incidents.

Normalization of deviance was initially described by Diane Vaughan (1996), a sociologist who studied the Challenger Space Shuttle explosion, as a long-term process in which individuals or teams accept a lower standard of performance until that lower standard becomes acceptable. This reflects an acceptance of risk shaped over time by deviant behavior that is never corrected because it is ignored or never detected. (Note that ignoring deviant behavior is a deviant behavior in its own right.) When enough deviant behavior has been ignored, it becomes accepted or normalized, from a sociological standpoint. Ultimately, no reaction is accepted as the appropriate reaction to deviant behavior, and the deviant behaviors themselves may come to be regarded as normal.

Normalization of deviance can be described in the context of the BP Texas City refinery explosion that caused 15 fatalities, 180 injuries, and 43,000 community citizens to be sheltered after evacuation. The refinery was starting up its isomerization unit when the raffinate splitter tower overfilled. Relief valves opened into an atmospheric blowdown stack. Flammable liquids then released from the stack, which led to an explosion and fire. The Baker Panel findings (Mufson, 2007, p. 15) cited BP as having no "common unifying process safety culture," a lack of operating discipline, and a toleration of serious deviation from safe operation practices leading to complacency around process safety risk. This normalization of deviance at BP was evident in the Baker Panel report: Operators accepted inoperable instrumentation, procedures were routinely not followed, operators were not adequately trained and the training staff was cut over time, nonessential workers were allowed to be near the process, and everyone accepted the faulty atmospheric blowdown stack arrangement.

Normalization of deviance happens frequently within any refining environment. Hazards are inherent in the processes of refining. People working in and around a refinery become so accustomed to being around these hazards that they become complacent in their inaction. For example, in any of the 139 refineries in the United States, one can observe the following scenario: An alarm sounds and people just turn it off. These alarms are connected to sophisticated pressure-monitoring systems that track chemicals as they travel through different parts of the refinery. Alarms are triggered whenever equipment experiences deviations beyond defined standards. The first time new workers experience a triggered alarm they tend to panic, but after a year, they hardly orient toward the console monitor before disabling the alarm. The management system failed to direct the mitigation behaviors it was designed to do. For example, forms used to inspect critical equipment can get pencil whipped (Ludwig, 2014), in which the inspector falsifies the form by filling it out without proper inspection. Safety equipment, such as eye wash and safety showers, can be left unmaintained. The list of management system failures can be extensive, resulting in most process safety risks unconsidered.

Consider the following example of process safety deterioration. At the IRD, there is a routine task of conducting surveys of pressure readings. While an operator was performing this task, he came across a bleeder line that was plugged. This bleeder line was designed to drain equipment and was not part of the pressure test. The operator decided to be proactive and unplug the bleeder, but he did not use the available bleeder tool because this tool had not been maintained and was no longer functional. He had witnessed most operators in the past use a simple screwdriver or wire to clear the bleeders and not experience a problem. In this case, the use of the wrong tool caused the plugged material of the bleeder to blow out and cause a vapor release. Fortunately, this vapor did not ignite; otherwise, the result would have been an explosion and fire. IRD's normalization of deviance was evident in that the proper tool was not maintained and the use of work-around tools was the norm.

Another incident occurred at the refinery coker. While an operator was making his normal rounds inspecting various equipment, he noticed a small vapor release from an insulated line near the thermowell. Operations decided to remove the insulation to get a better look at the leak, at which time inspectors were then contacted to investigate the issue. While they were inspecting the line, the leak worsened and a hydrocarbon vapor cloud filled a pump room. Fortunately, again, the vapor did not ignite, but this did cause an emergency unit shutdown.

Because these were significant near misses, causal analyses were conducted. The refinery personnel conducted what they considered a routine task but did not conduct a proper hazard recognition process. Improper tools were used for these tasks as well. Management systems also contributed to the near miss, in that leaders had not conducted an adequate risk-versus-reward analysis of the type of inspection mitigation the inspector performed. Because of this, small leaks were considered manageable, even though they actually represented a high risk. Indeed, smaller risk tasks require work permits, but this high-risk task did not. The normalization of deviance became clear: Pressure surveys became routine work, and the process of removing insulation became the norm. Management systems, such as hazard analyses and work permits, had not been updated because leaders had become complacent with the risk as well.

The oil and gas industry, along with its fellow industries in mining, food distribution, construction, and others, is actively seeking solutions to these behavioral challenges. With this in mind, it is clear that the behavior science community and its industry partners must build on what has been accomplished with personal safety and replicate and extend it to process safety. Industries need to replicate the use of behavior science, the identification of critical behaviors, the growth of a culture willing to report deviations, the development of a process-driven measurement system able to analyze deviance, and the ultimate reduction in the probability of incidents.

Managing to reduce normalization of deviance will take a greater understanding of the behavioral systems that lead to the actions associated with process safety risk. Adding behavior analyses to an understanding of normalization of deviance may help design new behavioral safety processes that can help reduce process safety incidents.

Many behaviors that deviate from the engineered work process are behaviors that simplify the work process (i.e., shortcuts). These shortcuts are directed through antecedents offered by the work context, the management systems, and sometimes the culture of the workforce (Mawhinney, 1992). For example, the work context and supervisory systems may create a perceived pressure to achieve a desired outcome based on past experiences or rumors. Management systems such as adequate training, proximal work instructions, and hazard analyses could be antecedents to direct proper behaviors. When these management systems are lacking sound behavioral processes, deviance occurs. Otherwise, poorly designed management systems, such as overly detailed and time-consuming work permit processes or lockout tagout procedures, may be antecedents for behaviors that result in less response cost. Finally, more simplistically, when appropriate tools are not available, the nearest tool becomes an antecedent itself. Over time, many of these antecedents become discriminative stimuli for shortcuts because of their association with reduction of response cost consequences.

There are often no immediate punishing consequences for engaging in the shortcut because the behavior rarely results in a process safety issue. In addition, because of their own normalization of deviance, supervisors do not discipline the operator for the shortcut. This makes the punishing consequence far less probable (Daniels & Bailey, 2014; Petrock, 1978), and the operator is rarely going to contact the punishing consequence. Instead, most shortcut behaviors (e.g., using the improper tool that is available, beginning work before completing a hazard form) will be reinforced because the response effort of the work is reduced, in that it is simpler, more convenient, or perhaps more physically comfortable, and the work gets done quicker, which may be further reinforced by supervisors or other employees.

In addition, behavioral systems analysis (Brethower, 1982; Diener, McGee, & Miguel, 2009; Harshbarger & Maley, 1974; Ludwig & Houmanfar, 2010; Malott, 2003) allows one to map how the action (e.g., improper tool use) or inaction (e.g., failing to conduct a hazard analysis) of the employee is occasioned by the interlocking contingencies (Glenn, 1988; Glenn & Malott, 2004; Malott & Glenn, 2006) surrounding the employee. Indeed, the behaviors of leaders fail to build systems that may set the context to direct and reinforce successful hazard mitigation behaviors of their employees. The behaviors of supervisors fail to provide more immediate direction and reinforcement. In fact, the behaviors of leaders may lead the supervisors to focus on other tasks (e.g., paperwork and meetings) at the expense of alternative behaviors directed at mitigating employee risk. In addition, the actions of people in other operational functions can impact the context of the operator (e.g., the warehouse function not procuring or maintaining the proper tools). Finally, one can assess how the behaviors of other operators who do the same task contribute to the behavioral drift that becomes the norm. It is clear that this drift can become a problematic, self-sustaining cycle.

The oil and gas industry is seeking the assistance of behavior science in studying, understanding, and mitigating these dramatic events associated with human behavior. The challenges for the behavior analysis community are to (a) research the behavioral root causes of normalization of deviance, (b) create behavioral interventions to reduce normalization of deviance, and (c) identify the system factors that would promote the behaviors necessary to avoid the development of normalization of deviance.

Basic behavior research can help understand this phenomenon, and applied research building practical behavioral methodologies can move these efforts forward. The oil and gas industry and MPC in particular have been successful in reducing personal injuries through hazard mitigation and risk reduction. The industry needs to move beyond an engineering approach to process safety and truly address its little understood behavioral components. The industry therefore calls on behavioral science researchers to study and understand the behaviors and reinforcement contexts that drive and maintain normalization of deviance, determine its root causes, and develop interventions to counteract the problem. MPC wants to use behavioral science to further its safety agenda, to ensure that workers go home uninjured, to keep host communities unaffected by negative externalities (Biglan, 2009), and to stop relying on luck. Personal and process safety hazards need to be studied and addressed. Researchers, practitioners, and industry employees must continue to collaborate in order to produce cutting-edge research that reduces human suffering.

REFERENCES

American Chemistry Council. (2014, July 8). *Responsible care.* Retrieved from http://responsiblecare.americanchemistry.com/?gclid=CjwKEAjwre6dBRC94d-Gma7g3wcSJACNatZesLbSepFHxbgPwDYYHDZZfloXF3CiUheZ5mON_urydxoCQqDw_wcB

Biglan, T. (2009). The role of advocacy organizations in reducing negative externalities. *Journal of Organizational Behavior Management*, *29*, 155–174.

Brethower, D. M. (1982). The total performance system. In R. M. O'Brien, A. M. Dickinson, & M. P. Rosow (Eds.), *Industrial behavior modification: A management handbook* (pp. 350–369). New York, NY: Pergamon Press.

Bureau of Labor Statistics. (2013, November 7). *How to compute a firm's incidence rate for safety management.* Retrieved from http://www.bls.gov/iif/osheval.htm

Cambridge Center for Behavioral Studies. (2014, July 8). *Companies achieving behavioral safety accreditation.* Retrieved from http://www.behavior.org/resource.php?id=327

Center for Chemical Process Safety. (2012, April 4). *Characteristics of the vapour cloud explosion incident at the IOC terminal in Jaipur, 29 October 2009.* Retrieved from http://www.aiche.org/ccps/resources/chemeondemand/conference-presentations/characteristics-vapour-cloud-explosion-incident-ioc-terminal-jaipur-29th-october-2009

Daniels, A., & Bailey, J. (2014). *Performance management: Changing behavior that drives organizational effectiveness* (5th ed.). Atlanta, GA: Performance Management.

Deepwater Horizon Study Group. (2011, March 1). *Final report on the investigation of the Macondo well blowout.* Retrieved from http://ccrm.berkeley.edu/pdfs_papers/bea_pdfs/dhsgfinalreport-march2011-tag.pdf

Deming, W. E. (1982). *Quality, productivity, and competitive position.* Cambridge: Massachusetts Institute of Technology, Center for Advanced Engineering Study.

Diener, L. H., McGee, H. M., & Miguel, C. (2009). An integrated approach to conducting a behavioral systems analysis. *Journal of Organizational Behavior Management*, *29*, 108–135.

Geller, E. S., Berry, T. D., Ludwig, T. D., Evans, R. E., Gilmore, M. R., & Clarke, S. W. (1990). A conceptual framework for developing and evaluating behavior change interventions for injury control. *Health Education Research: Theory and Practice*, *5*, 125–137.

Glenn, S. (1988). Contingencies and metacontingencies: Toward a synthesis of behavior analysis and cultural materialism. *The Behavior Analyst*, *11*, 161–179.

Glenn, S. S., & Malott, M. E. (2004). Complexity and selection: Implications for organizational change. *Behavior and Social Issues*, *13*, 89–106.

Harshbarger, D., & Maley, R. F. (1974). *Behavior analysis and systems analysis: An integrative approach to mental health programs.* Kalamazoo, MI: Behaviordelia.

Hermann, J. A., Ibarra, G. V., & Hopkins, B. L. (2010). A safety program that integrated behavior-based safety and traditional safety methods and its effects on injury rates of manufacturing workers. *Journal of Organizational Behavior Management*, *30*, 6–25.

Ludwig, T. D. (2014). The anatomy of pencil whipping. *Professional Safety*, *59*, 47–50.

Ludwig, T. D., & Geller, E. S. (2000). Intervening to improve the safety of delivery drivers: A systematic behavioral approach. *Journal of Organizational Behavior Management*, *19*(4), 1–124.

Ludwig, T. D., & Houmanfar, R. (Eds.). (2010). *Understanding complexity in organizations: Behavioral systems*. Philadelphia, PA: Routledge.

Malott, M. E. (2003). *Paradox of organizational change*. Reno, NV: Context Press.

Malott, M. E., & Glenn, S. S. (2006). Targets of intervention in cultural and behavioral change. *Behavior and Social Issues*, *15*, 31–56.

Mawhinney, T. C. (1992). Evolution of organizational cultures as selection by consequences: The Gaia hypothesis, metacontingencies, and organizational ecology. *Journal of Organizational Behavior Management*, *12*(2), 1–25.

McSween, T. E. (1995). *The values-based safety process: Improving your safety culture with a behavioral approach*. New York, NY: Wiley.

Mufson, S. (2007, January, 17). *BP failed on safety, report says*. Retrieved from http://www.washingtonpost.com/wp-dyn/content/article/2007/01/16/AR2007011600208.html

Myers, W. V., McSween, T. E., Medina, R. E., Rost, K., & Alvero, A. M. (2010). The implementation and maintenance of a behavioral safety process in a petroleum refinery. *Journal of Organization Behavior Management*, *30*, 285–307.

Occupational Safety and Health Administration. (2014a, April 30). *Industries in the VPP federal and state plans*. Retrieved from https://www.osha.gov/dcsp/vpp/sitebynaics.html

Occupational Safety and Health Administration. (2014b, July 8). *Process safety management*. Retrieved from https://www.osha.gov/SLTC/processs afetymanagement/

Petrock, F. (1978). Analyzing the balance of consequences for performance improvement. *Journal of Organizational Behavior Management*, *1*(3), 197–205.

Sulzer-Azaroff, B., & Austin, J. (2000). Does BBS work? Behavior-based safety and injury reduction: A survey of the evidence. *Professional Safety*, *45*, 19–24.

U.S. Chemical Safety and Hazard Investigation Board. (2008). *Investigation report: LPG fire at Valero-McKee refinery*. Retrieved from http://www.csb.gov/assets/1/19/CSBFinalReportValeroSunray.pdf

U.S. Chemical Safety and Hazard Investigation Board. (2009, July 19). *CITCO refinery hydrofluoric acid release and fire*. Retrieved from http://www.csb.gov/citgo-refinery-hydrofluoric-acid-release-and-fire/

U.S. Chemical Safety and Hazard Investigation Board. (2010). *Case study: Explosion and fire in West Carrollton, Ohio*. Retrieved from http://www.csb.gov/assets/1/19/Veolia_Case_Study.pdf

U.S. Chemical Safety and Hazard Investigation Board. (2013). *West Fertilizer explosion and fire*. Retrieved from http://www.csb.gov/west-fertilizer-explosion-and-fire-/

U.S. Chemical Safety and Hazard Investigation Board. (2014). *Investigation report: Catastrophic rupture of Heach exchanger*. Retrieved from http://www.csb.gov/assets/1/7/Tesoro_Anacortes_2014-May-01.pdf

Vaughan, D. (1996). *The Challenger launch decision: Risky technology, culture, and deviance at NASA*. Chicago, IL: University of Chicago Press.

Process Safety Behavioral Systems: Behaviors Interlock in Complex Metacontingencies

Timothy D. Ludwig

ABSTRACT

This paper seeks to identify behavioral components active in process safety. Three types of behavior classes are identified as contributors to process safety: task-specific behaviors, safety-directed behaviors, and behaviors associated with situational awareness. Behavioral systems analysis is used to provide a framework for identifying the cross-functional interlocking behavioral contingencies that can, even over a period of years, contribute to process safety incidents. Leadership behaviors are also identified that can create the context in the form of metacontingencies that maintain these interlocking contingencies.

The efficacious impact of behavioral approaches to reduce injury in industrial settings has been well documented. Initial studies (Fellner & Sulzer-Azaroff, 1984; Komaki, Barwick, & Scott, 1978) led to important foundational work (Alvero & Austin, 2004; Cooper, 2006, 200; Ludwig & Geller, 1997, 2000; Olson & Austin, 2001), reference books (Agnew & Daniels, 2010; Geller, 1996, 2002, 2005; McSween, 1995), and evidence-based case studies (Krause, Hidley, & Lareau, 1993; Myers, McSween, Medina, Rost, & Alvero, 2010; Ray, Purswell, & Bowen, 1991). Research in behavioral safety has been rooted in the science of behavior analysis and has grown into a globally-applied practice (Sulzer-Azaroff & Austin, 2000). In light of this success, industry leaders have called on the behavioral science community to provide the same rigor and expertise to understanding and impacting behaviors related to catastrophic incidents that kill, maim, pollute, and affect communities (Bogart, Ludwig, Staats, & Kretshmer, 2015).

In the modern media age, industrial disasters including Union Carbide's methyl isocyanate (MIC) release in Bhopal India (Broughton, 2005), the nuclear meltdown in Chernobyl, Ukraine (United Nations, 2002), the Exxon-Valdez oil spill (State of Alaska, 1990), and the NASA Challenger explosion (Vaughan, 1996) have been well-documented. Catastrophes like these have led to a discipline called process safety that refers to the

Color versions of one or more of the figures in the article can be found online at www.tandfonline.com/WORG.

management of hazards related to liquids and gasses that, through their toxic, reactive, or flammable qualities, can result in disaster (OSHA, 2000). Disasters that have focused attention on process safety include refinery explosions caused by companies such as British Petroleum (BP Texas City refinery explosion, Mufson, 2007), Valero (U.S. Chemical Safety and Hazard Investigation Board, 2008), Veolia (U.S. Chemical Safety and Hazard Investigation Board, 2010), and Tesoro (U.S. Chemical Safety and Hazard Investigation Board, 2014).

While its origins lie in the chemical and gas industries, process safety has been applied on a larger scale when referring to any industrial process where physical or chemical energies can be released to cause serious injuries across multiple people. This could include scaffolding failures in construction that kill workers and civilians (Kenny & Bracken, 2015; Occupational Safety and Health Administration [OSHA], 2010) and mining disasters (U.S. Department of Labor, 2010). Process safety management seeks to mitigate these significant hazards through technology, procedures, and management practices.

Process safety management has focused primarily on engineering and management processes to reduce the equipment failures that are associated with such disasters. OSHA published a *Process Safety Management-Guidelines for Compliance* (1994), which provides guidelines for process hazard analyses that OSHA (2000) summarizes as including, "engineering and administrative controls applicable to the hazards and their interrelationships, such as appropriate application of detection methodologies to provide early warning of releases." However, it is clear that human behavior interacts with these engineering processes; such behavior includes following engineered operating procedures, conducting equipment inspections, performing preventive maintenance, and implementing management systems (e.g., pre-startup safety reviews). Human behavior may allow equipment and safeguard controls to weaken through undesirable practices such as shortcuts, diverted attention, and leadership decisions that don't consider negative externalities (Bigland, 2009). Furthermore, this interaction between equipment failures and behavioral errors can often remain latent or undetectable until they cascade into fatal consequences as described above. Indeed, OSHA (2000) also calls for analyses of "human factors" (c.f., Rodriguez, et. al., 2017), the overriding term for managing the behavior of individuals in a system, as part of their guidelines for process safety management.

The 2010 explosion of Deepwater Horizon in the Gulf of Mexico exemplifies how human behavior can interact with critical equipment and processes to disastrous ends. Many equipment failures that led to this disaster can be linked to human behavior (National Commission on the BP Deepwater Horizon Oil Spill and Offshore Drilling, 2011; U.S. Chemical Safety and Hazard Identification Board, 2014). One such equipment failure involved blowout preventers, which are clamps that sever and seal the drill piping if upward pressure from the well is not controlled. When BP lost control of the well, these blowout preventers failed to work. The resulting investigation concluded two separate incidents of miswiring by electricians many months before deployment that led to battery failure within the blowout preventers. This problem was further compounded by the omission of required inspections, allowing this latent failure to manifest.

In the midst of these equipment failures were fatal management behaviors in the form of decisions and direction. Crew testimonies recount an angry interchange between the Transocean rig manager and the BP manager (Barstow, Rohde, & Saul, 2010). The rig manager, knowing that the well had more upward pressure than desired, wanted to use a cautious well casing (cementing) procedure that would keep adequate downward pressure. Federal guidelines required the more cautious procedure with the pressure levels. The crew even suspected that the blowout preventers had failed when they saw plastics coming out of the "mud" used to maintain downward pressure. Knowing that this would result in millions of dollars in added time, the company manager reportedly directed the rig manager to use a quicker process. The quicker process of casing the well, which was executed by a contractor without objection from its management, resulted in the subsequent release and explosion.

The offshore drilling rig had established and practiced procedures to assure all on-board had access to lifeboats in the event of an emergency. However, these procedures were abandoned during the chaos of the explosion and resulting fire, and many employees had no lifeboats to carry them to safety. The accounts of having to jump off the 10-story platform into burning oil on the surface of the sea are harrowing (Barstow, 2010). The result of the cascading series of events involving behaviors of numerous actors across time, companies, and functions left 11 people dead and caused ruinous environmental damage, costing BP tens of billions of dollars.

This paper seeks to take a first step in identifying the behavioral system components active in process safety toward the ultimate goal of providing a scientific framework that may generate applied efforts to build interventions and systems that contribute to process safety management.

Components within behavioral systems contributing to process safety

Austin (2000), Daniels and Bailey (2014), Diener, McGee, and Miguel (2009), and Malott (2003) all offered Organizational Behavior Management (OBM) methodologies to "pinpoint" the behaviors and response classes that would be most associated with performance. Pinpointing is the process of identifying behaviors related to performance variables being analyzed. Hyten (2002) argues that behaviors targeted in applied research, as well as in practice, must be related to meaningful organizational results. The behavior of workers that could either prevent or proactively identify process safety issues might include classes of behavior such as:

(a) On-task behaviors (operants) that can either strengthen or weaken engineered controls. Behaviors that vary from operating procedures (e.g., engaging in shortcuts, using improper tools; cf. Bogard et al., 2015) can degrade equipment during manufacturing, setup, operation, and maintenance such as when electrician behaviors lead to inadequate wiring that cause battery failure.
(b) Safety-directed behaviors that are either operants (e.g., lockout/tagout) or classes of verbal behavior. These are behaviors that are involved in reviewing job procedures in job safety analyses (JSAs), conducting inspections, and hazard identification activities that occur before engaging in task behaviors. These could include omission of behaviors such as responding to alarms (Bogard et al., 2015), conducting thorough inspections, or preventative maintenance. This may unfortunately include behaviors involved in "pencil whipping" (Ludwig, 2014) safety management systems involved in inspections, permitting, or reporting, leaving the false impression that procedures had been followed while leaving a latent hazard undetected.
(c) A final class related to "situational awareness" is shaped when environmental stimuli elicit stimulus control processes that might lead to scanning, checking, referring to instructions, and other overt behaviors (Hogan, Bell, & Olson, 2009; Killingsworth, Miller, & Alavosius, 2016). For example, painters on scaffolding among refinery pipes can conduct visual scans of the pipes and joints for corrosion and damage while doing their job.

Process safety research and practice should, therefore, consider many broad classes of behavior in addition to solely focusing on operating procedure compliance.

Even meaningful pinpoints, often targeting one job task, can fall short of describing the sometimes complex interlocking behaviors across many agents (Mihalic & Ludwig, 2009), in cross-functional metacontingencies (Glenn, 1988; Glenn & Malott, 2004; Malott & Glenn, 2006), that function to manage controls in process safety or, in contrast, create conditions or behavioral variance (Ludwig, 2002) that raise the risk level for a process incident.

Behaviors emitted by individual agents throughout the system by operators, maintenance, contractors, managers, engineers, planners, leaders, and other support staff (information technology, human resources, procurement, etc.) influence both the behavior and the outcomes of behavior by other agents of the system. In these metacontingencies, behaviors of one individual in the organization (e.g., manager) serve as antecedents and consequences for others' behavior (Blasingame & Clayton, Mawhinney, Luke, & Cook, 1997; Glenn, 1988, 1991; Ludwig, 2014). A simple example is how a planner's behavior that sets out the sequence of tasks to be performed on a maintenance activity in turn creates the antecedents directing the work for a maintenance employee. Reciprocally, the resulting behavior of the receiving agent (maintenance employee) creates outcomes that reinforce or punish the behaviors of the originating agent (the planner). If the planner's plan underestimated the time needed to complete the task (an antecedent), this may result in modifications of the maintenance employee's behavior to produce the timeliness outcome regardless of shortcuts to the operating procedure. Completing the task within the strict timeline could, in turn, be a reinforcer for the planner's scheduling behavior, thereby producing dangerously tight timelines in the future.

In more complex processes, there are likely to be many more than two agents whose behavior impacts process controls. For each additional agent in the metacontingency, the number of interlocked behavioral contingencies (IBCs) increases exponentially, adding to its complexity (Glenn & Malott, 2004). All of these distinct yet connected agent behaviors would be contributing to the metacontingency resulting in greater productivity, further reinforcing these interlocking behaviors. The IBCs continue to maintain individual behavior and, as a composite, a site's ability to maintain control over its process safety. Alternatively, they are more likely to be sources of behavioral variance that drive key pinpoints out of behavioral control limits (Hyten & Ludwig, 2017) within operating procedures, process safety management systems, or leadership decision making. When this happens the risk of process safety incidents greatly increase.

Over time, these reciprocal consequences will further adapt agent behaviors, thereby creating a milieu of group behavior that can be hard to change (Clayton et al., 1997). Because of the complexity and the incremental nature of changes occasioned by the IBCs, the agents themselves may not be aware of the resulting increased behavioral variance. This could be a contributing factor to the phenomena called normalization of deviance (Bogard et al., 2015; Vaughan, 1996). It is difficult enough to recognize variance in one's own behavior over time; it would be exponentially more difficult to be aware of variance when each agent's behavior interlocks in ways that incrementally adjust social and operational contingencies. This would create variance in both individual behaviors and the system itself; additionally, this would normalize the variance because the new behavioral contingencies are the norm.

Thus, there is a striking need to define the IBCs, what they produce, and who receives these products of behavior (Glenn & Malott, 2004) in order to better understand and operate within these often-complex metacontingencies of process safety. To break through the complexity of organizational metacontingencies, Ludwig and Houmanfar (2010), Malott (2003), McGee and Diener (2010), 2012), and Hyten (2009), among others, argue that an organization's business environment, internal system, processes, and feedback loops are the strongest context of metacontingencies, subsequent interlocking contingencies, and local behavioral contingencies. Therefore, system assessment (Hyten, 2009; McGee & Diener-Ludwig, 2012; Rummler & Brache, 1995) may be a first step in the assessment and improvement of human factors within process safety.

Behavioral Systems Analysis (BSA; Brethower, 1982; Diener et al., 2009; Harshbarger & Maley, 1974; Ludwig & Houmanfar, 2010; Malott, 2003) provides a framework to consider the full range of systems variables before targeting behaviors for more meaningful change. Ludwig and Houmanfar (2010) introduced BSA in special issues of the *Journal of Organizational Behavior Management* explaining that,

> systems are adaptive entities that survive by meeting environmental demands (consumers, competition, economy, governmental policies, etc.) through the development and maintenance of subsystems ultimately designed to manage behavior. Thus, organizations are behavioral systems that encompass complex patterns of behavioral interactions among its members and the environment. (p. 85)

BSA seeks to understand these complex response patterns in the context of critical components of the organizational system to better manipulate the relationships between the organization and behavioral contingencies to achieve desired organizational outcomes (e.g., process safety controls).

A model created by Rummler (Rummler & Brache, 1995; Rummler, 2001; 2004) described three levels within organizations where systemic relationships exist: (a) an organization level that incorporates the company goals or strategy, (b) a process level such as workflow or information flow across functional departments that make up the organization, and (c) a job/performer level focusing on the behavior of individual employees and managers. The organizational level of performance emphasizes the organization's relationship with outside environments, internal structure, and distribution of resources. These macro-level factors set the context for the shared metacontingency outcomes derived from the aggregate products of interlocked behaviors in the behavioral system. Thus, many involved in the application of behavioral science to process safety (Hyten & Ludwig, 2017; McSween & Moran, 2017; Rodriguez et al., 2017) have argued for interventions at the board level of for-profit companies, whereby the board exerts control over the metacontingencies and, therefore, over the critical decision making of leaders that impacts not only the resources dedicated to process safety, but also the cascading contingencies (i.e., Consequence Chain; Gravina, et al., 2017) throughout the organization.

For more practical applications of BSA to process safety we should drill down to defining the more direct IBCs occasioned by the organizational structure and processing systems surrounding our process safety pinpoints. The processing system focuses on the internal design and structure of the organization that contributes to the throughput of the organization (Rummler & Brache, 1995). In this regard we seek to understand how departmental agents interact to complete work and identify where these interactions may be flawed, thereby producing metacontingencies that create the context for sub-optimal process safety performance. One useful BSA tool is the relationship map (Rummler & Brache, 1995) that seeks to describe the relationships between different functions (i.e., a business unit with its own reporting structure organized around an operational or supporting output), both inside and outside of the organization.

Figure 1 depicts a probable behavioral system's relationship map (Rummler & Brache, 1995) whose actors engage in behaviors that interlock, within metacontingencies, with other agents of the organization and its affiliates, all of whom engage in the three classes of behavior described (i.e., on-task, safety directed, and situational); the product of these metacontingencies is the maintained control of process safety. Note the numerous components to the network of relationships that can impact a process safety outcome. Functions are depicted in boxes. Functions within the wide box represent functions within a fictional petrochemical company. These may include operations, warehouse and maintenance, support functions such as engineering and safety, as well as the plant leadership. Supplier functions from outside the company are depicted on the left. Finally, external actors

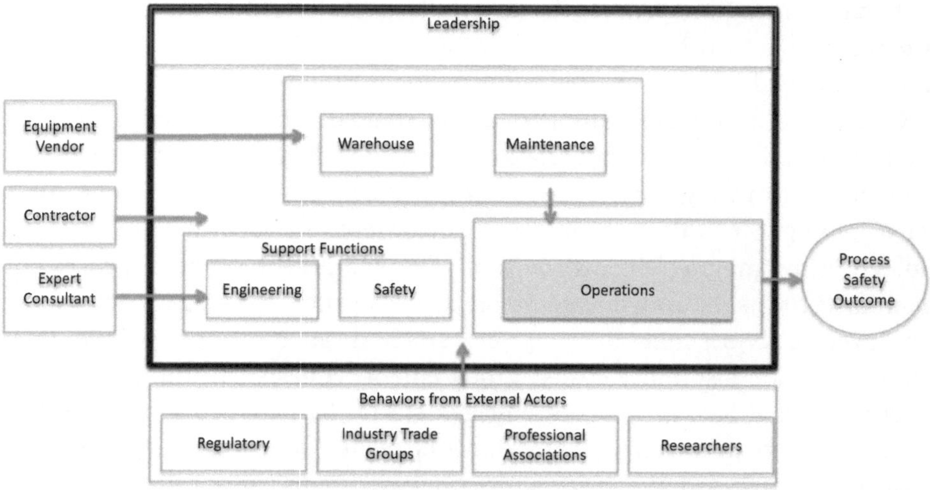

Figure 1. An example of a relationship map depicting the different functions of an industrial project.

who function to impact actions of the plant are shown below the company. The process safety outcome is depicted with the circle coming out of this system to the right.

Example of interlocking behavioral contingencies in a process safety close call

Workers on a team may engage in behaviors through work processes that are inconsistent with operating procedures and produce variance creating an emerging latent hazard. Subsequently, other team members unaware of prior variance might, through their own behavior, trigger the latent hazard increasing the probability of an incident.

Workers in other functions also engage in interlocking behaviors that may result in a causal chain of events creating the context that the two workers described above work within. These behaviors are the precursors to the events that create the latent hazard or leave one undetected. For example, maintenance staff are constantly interacting with equipment and engineering controls through ongoing preventive maintenance, repair, and installation work. Inventory (warehouse) staff provide tools and parts that need to be readily available and fit for purpose. Internal functional "silos" with different work processes and management chains may lead to a lack of coordination of the interlocking behaviors necessary to maintain the integrity of the controls put in place for process safety. Further, contractors hired to provide services like electrical and plumbing work, scaffolding, painting, insulation, and grading all engage in behaviors that interlock with multiple agent behaviors (internal and external) as they interact with operational equipment and processes. The fact that contractors tend to be external agents working under different organizational contingencies can cause these IBCs to be overlooked or undermanaged.

Consider a simplified example of a non-standard casing causing a process safety close call displayed in a relationship map in Figure 2. Many years back engineers retrofitting part of the plant designed a nonstandard casing for a set of piping. Since these casings are not ordered frequently, they are considered "slow moving" in distribution. The equipment vendors' process of manufacturing slow moving items on a just-in-time basis result in vendor employee behaviors that fail to deliver parts in a timely manner (Circle "A"). These items are also more costly. Procurement requests left unfulfilled seem to punish warehouse employees' engagement in the paperwork because they begin to avoid time-consuming behaviors related to writing up replenishment procurement requests for slower moving items, only procuring items that get used frequently. This also helps the warehouse budget by saving money on parts rarely needed.

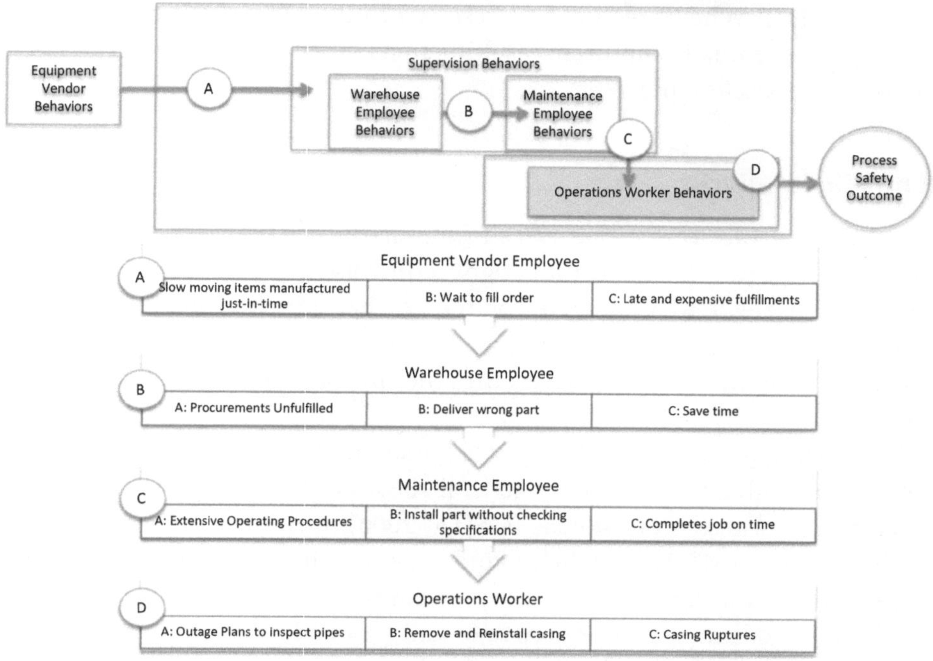

Figure 2. Interlocking behavioral contingencies between equipment vendors, warehouse and maintenance employees, impacting the product of operation worker behaviors. Circled letters represent events listed in the text. Within the table, "A" is the antecedent, "B" is the behavior, and "C" is the consequence for the agent listed.

In the future, a maintenance agent requested a casing from the warehouse consistent with the operating procedures for the equipment (Circle "B"). However, that specific size casing had not been replenished due to the above IBC. Unaware of the process safety specification on the operating procedures, a warehouse agent provided an available and functional, but incorrect, casing to maintenance. Doing this may allow the warehouse agent to avoid potential negative consequences for the omission of procurement requests revealed by the casing being out of stock. Procuring the item at this stage would require an extensive time delay on the planned task, further increasing the probability of negative consequences

The maintenance agent, working through an extensive operating procedure whose planner underestimated the time budget, omitted situational awareness behaviors needed to check individual parts to assure they conformed to their specification. The maintenance agent completes the preventive maintenance job and fits the incorrect casing on the piping (Circle "C").

Many years pass where this casing remains in place without failing. A future operator assumed all was in order and, during an outage targeting this piece of equipment, removed the casing to inspect the pipe before reinstalling it (Circle "D"). When pressure was restored to the piping, the casing began leaking and eventually ruptured, resulting in a release and, fortunately, only a minor process safety incident.

Heuristic interlocks

Regardless of the source, the IBCs of hands-on behaviors can either lead to the successful execution of tight controls on process hazards or, alternatively, lead to undetected variance that can become common in work practices (i.e., normalization of deviance; Bogard et al., 2015; Vaughan, 1996), thereby increasing the probability of process safety incidents.

In the United States, OSHA provides guidelines for process safety management, which attempt to pinpoint general classes of behaviors and behavioral outcomes for safety professionals (OSHA, 2000), including the management of "human factors." Thus, OSHA guidelines require that employers put in place methods through which employees can participate in the "conduct and development of process hazard analyses and on the development of the other elements of process management ... " (OSHA, 2000).

Those tasked with monitoring and managing process safety through a more heuristic class of verbal behaviors involved in decision making are a professional group of employees, whose pay tends to be salaried, that provide the design and management systems that direct and enable the interlocking behaviors of the workforce. Safety professionals engage in a series of behaviors to benchmark, design, implement, support, and review safety management systems such as processes/procedures, hazard identification tools, behavioral safety, audits/inspections, etc.; all of which produce data to be analyzed and acted upon. This feedback loop after incident investigations or, more proactively with leading indicator data, is a class of behaviors that help identify potential risks and hazards so that corrective actions can be developed, implemented, and evaluated.

Engineers who design the equipment and controls central to process safety are an obvious part of the behavioral system. Engineers not only engage in a series of behaviors that design equipment and controls, they also work on teams to review designs to discover flaws, omissions, and potential user issues that may lead to on-the-ground behavioral variance due to complexity, difficulty, or incompatibility. Engineer behaviors result in equipment changes, manuals, operating procedures, labeling, instructions, and management of change actions; all of which serve as antecedents to the hands-on behaviors. Engineering outcomes should therefore be built with better behavior analytic precision beyond the technical steps, to assure more fluent execution by workers. When engineers behave in ways that take them out to the field to observe worker interactions with their designs and instructions, they provide themselves with the necessary inputs to modify their own behavior to this end.

The behaviors from the managerial group of employees can have even more profound impact on the metacontingencies present in these behavioral systems; in some cases, such as within leadership behaviors, individuals can modify these metacontingencies. Leaders' verbal behavior set up establishing operations (EOs) when they describe the values of the organizational division, while their overt behaviors evident in their ongoing verbalizations (vocal and electronic) and decisions either strengthen or weaken the EOs (Houmanfar, Rodrigues, & Smith, 2009). Leaders' decisions and follow-up through the IBCs with their subordinates build the contingencies that either strengthen multiple agent behaviors toward process safety outcomes or degrade them by directing behavior elsewhere. Similarly, the behavior of the supervisory staff in direct contact with the worker proximal to the process safety controls has the same, albeit more direct, effect on the contingencies maintaining the behavioral classes impacting process safety. Specific supervisor behaviors to pinpoint would include providing instructions, approvals, oversight, and feedback as well as planning, securing needed tools and repairs, and handing off a task from one worker or workgroup to another.

We continue our example of the casing leak in Figure 3. Prior to the events discussed thus far, a top leader set in motion a series of metacontingencies through his ongoing verbal behaviors promoting cost reductions, overtly linking them to higher performance ratings and bonuses for managerial staff who succeeded in coming in under budget (Circle "A_1"). This contingency acts on the behavior of supervisors who may start denying or delaying requisition requests of slow-moving warehouse items ("A_2"), which in turn keeps vendors from providing the needed parts, thereby setting off the chain of interlocking worker behaviors related to the incident ("B, C, and D").

Environmental interlocks

Behaviors of individuals in the super system (Rummler & Brache, 1995), processes outside the organizational system that impact its functioning, can have either a direct or more nuanced effect on the metacontingencies surrounding process safety. Figure 1 listed some of these entities. Government regulators who research and write guidelines (e.g., OSHA, 2000) or conduct audits and on-site inspections do so to compel the behaviors of those in an organization directly through the threat of sanctions. Professional associations such as the American Society of Safety Engineers (ASSE.org) provide the context for members to behave in ways to share best practices. Consultant behaviors play a role in delivering programs and training that shape behaviors, while researcher behaviors use publications such as this to grapple with basic questions in scientific fields, in order to offer guidance to those practicing their craft and to reduce catastrophic events through process safety management.

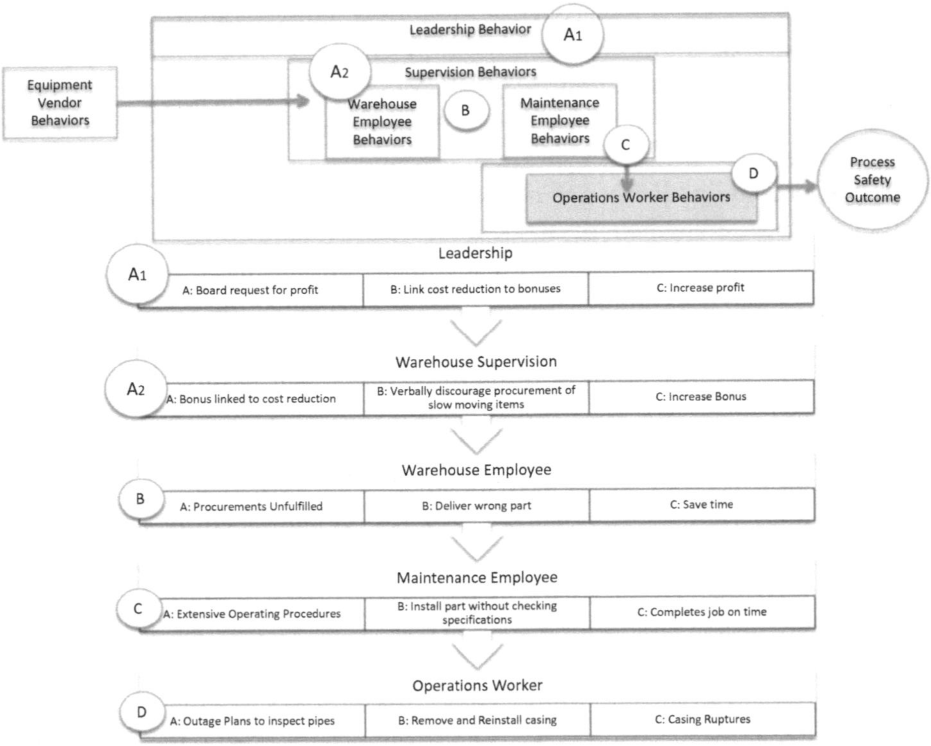

Figure 3. Metacontingencies set in motion by leadership behaviors cascading to supervisory behaviors that impact the interlocking contingencies between employees across different functions. Circled letters represent events listed in the text. Within the table, "A" is the antecedent, "B" is the behavior, and "C" is the consequence for the agent listed.

Conclusion

The IBCs representing the myriad of individuals acting within a complex integrated behavioral system must be understood and mapped to effectively build process safety management tactics to standardize, design, review, build, inspect, operate, evaluate, and improve the equipment, processes, and human relationships related to process safety. Assessments of process safety behavioral systems must be developed for organizations to assess their systemic vulnerabilities across functions to reveal what may be latent and normalized. Any analysis or implementation of technical or behavioral tactics designed to control process safety variables must consider how the behavior of individuals in these different roles have a sustained impact on the resulting variance across the behavioral system; variance related to latent hazards, normalization of deviance, and countercontrol discovered only after a (sometimes catastrophic) process safety incident.

References

Agnew, J. L., & Daniels, A. C. (2010). *Safe by accident?* Atlanta, GA: Performance Management.

Alvero, A. M., & Austin, J. (2004). The effect of conducting behavioral observations on the behavior of the observer. *Journal of Applied Behavior Analysis, 37*, 457–468. doi:10.1901/jaba.2004.37-457

Austin, J. (2000). Performance analysis and performance diagnostics. In J. Austin, & J. Carr (Eds.), *Handbook of applied behavior analysis* (pp. 304–327). Reno, NV: Context Press.

Barstow, D., Rohde, D., & Saul, S. (2010, December 25). Deepwater horizon's final hours. *The New York Times*.

Biglan, T. (2009). The role of advocacy organizations in reducing negative externalities. *Journal of Organizational Behavior Management, 29*, 155–174. doi:10.1080/01608060903092086

Bogard, K., Ludwig, T. D., Staats, C., & Kretchmer, D. (2015). An industry's call to understand the contingencies involved in process safety: Normalization of deviance and interlocking contingencies. *Journal of Organizational Behavior Management, 35*, 70–80. doi:10.1080/01608061.2015.1031429

Brethower, D. M. (1982). The total performance system. In R. M. O'Brien, A. M. Dickinson, & M. P. Rosow (Eds.), *Industrial behavior modification: A management handbook* (pp. 350–369). New York, NY: Pergamon Press.

Broughton, E. (2005). The Bhopal disaster and its aftermath: A review. *Environmental Health, 4*, 6. doi:10.1186/1476-069X-4-6

Clayton, M. C., Mawhinney, T. C., Luke, D. E., & Cook, H. G. (1997). Improving the management of overtime costs through decentralized controls: Managing an organizational metacontingency. *Journal of Organizational Behavior Management, 17*(3), 77–97. doi:10.1300/J075v17n02_03

Cooper, D. M. (2006). Exploratory analyses of the effects of managerial support and feedback consequences on behavioral safety maintenance. *Journal of Organization Behavior Management, 26*, 1–41. doi:10.1300/J075v26n03_01

Cooper, M. D. (2009). Behavioral Safety Interventions: A review of process design factors. *Professional Safety, 54*(2), 36.

Daniels, A., & Bailey, J. (2014). *Performance management: Changing behavior that drives organizational effectiveness* (5th ed.). Atlanta, GA: Performance Management.

Diener, L. H., McGee, H. M., & Miguel, C. (2009). An integrated approach to conducting a behavioral systems analysis. *Journal of Organizational Behavior Management, 29*, 108–135. doi:10.1080/01608060902874534

Fellner, D. J., & Sulzer-Azaroff, B. (1984). Increasing industrial safety practices and conditions through posted feedback. *Journal of Safety Research, 15*, 17–21. doi:10.1016/0022-4375(84)90026-4

Geller, E. S. (1996). *The psychology of safety*. Randor, PA: Chilton Book Company.

Geller, E. S. (2002). *The participation factor: How to increase involvement in occupational safety*. Des Plaines, IL: American Society of Safety Engineers.

Geller, E. S. (2005). *People-based safety: The source*. Virginia Beach, VA: Coastal Training Technologies Corporation.

Glenn, S. (1988). Contingencies and metacontingencies: Toward a synthesis of behavior analysis and cultural materialism. *The Behavior Analyst, 11*, 161–179. doi:10.1007/BF03392470

Glenn, S. (1991). Contingencies and metacontingencies: Relations among behavioral, cultural, and biological evolution. In P. A. Lamal (Ed.), *Behavioral analysis of societies and cultural practices* (pp. 39–73). New York, NY: Hemisphere Publishing Corporation.

Glenn, S. S., & Malott, M. E. (2004). Complexity and selection: Implications for organizational change. *Behavior and Social Issues, 13*, 89–106. doi:10.5210/bsi.v13i2.378

Gravina, N., Cummins, B., & Austin, J. (2017). Leadership's role in process safety: An understanding of Behavioral science among managers and executives is needed. *Journal of Organizational Behavior Management, 37*(3–4), 316–331.

Harshbarger, D., & Maley, R. F. (1974). *Behavior analysis and systems analysis: An integrative approach to mental health programs.* Kalamazoo, MI: Behaviordelia.

Hogan, L. C., Bell, M., & Olson, R. (2009). A preliminary investigation of the reinforcement function of signal detections in simulated baggage screening: Further support for the vigilance reinforcement hypothesis. *Journal of Organizational Behavior Management, 29* (1), 6–18. doi:10.1080/01608060802660116

Houmanfar, R. A., Rodrigues, N. J., & Smith, G. S. (2009). Role of communication networks in behavioral systems analysis. *Journal of Organizational Behavior Management, 29*, 257–275. doi:10.1080/01608060903092102

Hyten, C. (2002). On the identity crisis in OBM. *The Behavior Analyst, 3*(3), 301–310. doi:10.1037/h0099982

Hyten, C. (2009). Strengthening the focus on business results: The need for systems approaches in OBM. *Journal of Organizational Behavior Management, 29*(2), 87–107. doi:10.1080/01608060902874526

Hyten, C., & Ludwig, T. (2017). Complacency in process safety: A Behavior analysis toward prevention strategies. *Journal of Organizational Behavior Management, 37*(3–4), 240–260.

Kenney, A., & Bracken, D. (2015). Raleigh scaffolding collapse kills 3 construction workers. *The News Observer.* Retrieved October 13, 2015, from http://www.newsobserver.com/news/local/counties/wake-county/article16081955.html

Killingsworth, K., Miller, S. A., & Alavosius, M. P. (2016). A behavioral interpretation of situation awareness: Prospects for OBM. *Journal of Organizational Behavior Management, 36*, 301–321. doi:10.1080/01608061.2016.1236056

Komaki, J., Barwick, K. D., & Scott, L. R. (1978). A behavioral approach to occupational safety: Pinpointing and reinforcing safe performance in a food manufacturing plant. *Journal of Applied Behavior Analysis, 63*, 434–445.

Krause, T. R., Hidley, J. H., & Lareau, W. (1993). Implementing the behavior-based safety process in a union environment: A natural fit. *Professional Safety, 38*(6), 26–31.

Ludwig, T. D. (2002). On the necessity of structure in an arbitrary world: Using concurrent schedules of reinforcement to describe response generalization. *Journal of Organizational Behavior Management, 21*(4), 13–38. doi:10.1300/J075v21n04_03

Ludwig, T. D. (2014). The anatomy of pencil whipping. *Professional Safety, 59*, 47–50.

Ludwig, T. D., & Geller, E. S. (1997). Assigned versus participatory goal setting and response generalization: Managing injury control among professional pizza deliverers. *Journal of Applied Psychology, 82*(2), 253–261. doi:10.1037/0021-9010.82.2.253

Ludwig, T. D., & Geller, E. S. (2000). Intervening to improve the safety of delivery drivers: A systematic behavioral approach [Monograph]. *Journal of Organizational Behavior Management, 19*, 1–153. doi:10.1300/J075v19n04_01

Ludwig, T. D., & Houmanfar, R. (Eds.). (2010). *Understanding complexity in organizations: Behavioral systems.* Philadelphia, PA: Routledge.

Malott, M. E. (2003). *Paradox of organizational change.* Reno, NV: Context Press.

Malott, M. E., & Glenn, S. S. (2006). Targets of intervention in cultural and behavioral change. *Behavior and Social Issues, 15*, 31–56. doi:10.5210/bsi.v15i1.344

McGee, H. M., & Diener, L. H. (2010). Behavioral systems analysis in health and human services. *Behavior Modification, 34*, 415–442. doi:10.1177/0145445510383527

McGee, H. M., & Diener-Ludwig, L. H. (2012). An introduction to behavioral systems analysis for rehabilitation agencies. *Journal of Rehabilitation Administration, 36*(2), 59–71.

McSween, T. E. (1995). *The values-based safety process: Improving your safety culture with a behavioral approach*. New York, NY: Wiley.

McSween, T., & Moran, D. (2017). Assessing and preventing serious Incidents with behavioral science. *Journal of Organizational Behavior Management, 37*(3–4), 283–300.

Mihalic, M. T., & Ludwig, T. D. (2009). Behavioral system feedback measurement failure: Sweeping Quality under the rug. *Journal of Organizational Behavior Management, 29*(2), 155–174. doi:10.1080/01608060902874559

Mufson, S. (2007, January, 17). BP failed on safety, report says. Retrieved from http://www.washingtonpost.com/wp-dyn/content/article/2007/01/16/AR2007011600208.html

Myers, W. V., McSween, T. E., Medina, R. E., Rost, K., & Alvero, A. M. (2010). The implementation and maintenance of a behavioral safety process in a petroleum refinery. *Journal of Organization Behavior Management, 30*, 285–307. doi:10.1080/01608061.2010.499027

National Commission on the BP Deepwater Horizon Oil Spill and Offshore Drilling. (2011). *Deep water: The gulf oil disaster and the future of offshore drilling*. Washington, DC: U.S. Government Printing Office.

Occupational Safety and Health Administration (1994). *Process safety management guidelines for compliance*. Retrieved May 13, 2016, from https://www.osha.gov/Publications/osha3133.html

Occupational Safety and Health Administration (2000). *Process safety management*. Retrieved May 13, 2016, from https://www.osha.gov/Publications/osha3132.html

Occupational Safety and Health Administration (2010, November 2).*U.S. labor department's OSHA cites New York and Pennsylvania contractors following scaffold collapse at Binghamton University that injured 6 workers*. Retrieved October 13, 2015, from https://www.osha.gov/pls/oshaweb/owadisp.show_document?p_table=NEWS_RELEASES&p_id=18621

Olson, R., & Austin, J. (2001). Behavior-based safety and working alone: The effects of a self-monitoring package on the safe performance of bus operators. *Journal of Organizational Behavior Management, 21*(3), 5-43. doi:10.1300/J075v21n03_02

Ray, P. S., Purswell, J. L., & Bowen, D. J. (1991). Long-term effect of a behavioral safety program. In W. Karwowski, & J. W. Yates (Eds.), *Advances in industrial ergonomics and safety* (pp. 725–730). London, UK: Taylor & Francis.

Rodriguez, M. A., Bell. J., Brown. M., & Carter. D. (2017). Integrating Behavioral Science with Human Factors to Address Process Safety. *Journal of Organizational Behavior Management, 37*(3–4), 301–315

Rummler, G. A. (2001). Performance logic: The organization performance Rosetta stone. In L. J. Hayes, J. Austin, R. Houmanfar, & M. C. Clayton (Eds.), *Organizational change* (pp. 111–132). Reno, NV: Context Press.

Rummler, G. A. (2004). Serious performance consulting according to Rummler. Silver Spring, MD: International Society for Performance Improvement.

Rummler, G. A., & Brache, A. P. (1995). *Improving performance: How to manage the white space on the organizational chart*. San Francisco, CA: Jossey-Bass.

State of Alaska (1990). *Final report, alaska oil spill commission. http://www.evostc.state.ak.us/index.cfm?FA=facts.details*

Sulzer-Azaroff, B., & Austin, J. (2000). Does BBS work? Behavior-based safety & injury reduction: A survey of the evidence. *Professional Safety, 45*, 19–24.

U.S. Chemical Safety & Hazard Identification Board (2008). LPG Fire at Valero-McKee Refinery. REPORT NO. 2007-05-I-TX. http://www.csb.gov/file.aspx?DocumentId=421

U.S. Chemical Safety & Hazard Identification Board (2010). Explosion and Fire in West Carrollton, Ohio. REPORT NO. 2009-10-I-OH. http://www.csb.gov/file.aspx?DocumentId=425

U.S. Chemical Safety and Hazard Identification Board (2014). Explosion and fire at the macondo well. Report No. 2010-10-I-OS. Retrieved from http://www.csb.gov/assets/1/7/Overview_-_Final.pdf

United Nations (2002). The human consequences of the chernobyl nuclear accident. Retrieved from http://www.un.org/ha/chernobyl/docs/report.pdf

United States Department of Labor, Mine Safety and Health Administration (April 2010). *Report of investigation: Fatal underground mine explosion upper big branch mine-south, performance coal company*. Retrieved from http://www.msha.gov/Fatals/2010/UBB/FTL10c0331.pdf

Vaughan, D. (1996). *The Challenger launch decision: Risky technology, culture, and deviance at NASA*. Chicago, IL: University of Chicago Press.

Complacency in Process Safety: A Behavior Analysis Toward Prevention Strategies

Cloyd Hyten and Timothy D. Ludwig

ABSTRACT

Complacency inhibits safe behaviors of workers and managers. This is of concern to industries where process safety is needed to reduce the chance of catastrophic events such as fires and explosions. A behavioral definition of complacency is offered as trending behavioral variation that eventually exceeds safety boundaries. Behavioral processes that contribute to these patterns of variability are discussed and analyzed, including habituation, extinction, unprogrammed reinforcement, the avoidance paradox, rule-governed behavior, and competing contingencies of production. Solution strategies are suggested that address this analysis of behavioral variance, including pinpointing behavioral variation related to safety, changing training design, strengthening positive reinforcement for process-related behaviors of workers and management, reducing sources of unprogrammed reinforcement for dangerous variation, strengthening rule-governed behavior, and changing contingencies for managers and executives whose decisions affect behavior and process safety at many levels in the company.

A process safety incident can be defined as an unplanned event arising from the manufacturing process that results in a product spill, fire, explosion, or injury (American Chemistry Council, 2017; OSHA, 2000). A root cause analysis is typically a major part of the aftermath of a process safety incident or near miss. Root cause analyses seek to find the underlying origin of the chain of events that resulted in the incident. These may include equipment failure, environmental events such as inclement weather, or work process deviations within or across functions (Ludwig, 2017).

When root cause analyses detect work process deviations due to human error, it is often implied that the person had become "complacent," seen as a personal failing in the sense that that the person let themselves become complacent through some form of neglect. Thus, it is common for management to impose remedies such as re-training or discipline to jolt the problematic worker out of this complacent state (Carroll, Rudolph, &

Hatakenaka, 2002). This is a mistake on several levels: (a) deficient behavior should not be seen as a root cause because the causes of these behaviors still need to be identified, in other words, the causes of the error and the complacency have yet to be explained; (b) blaming the worker ignores the role that management practices and systems played in the incident; therefore, (c) the solutions may not address the real causes of the problem, wasting time, money, and failing to reduce the risk of subsequent process safety incidents or personal injury (Hyten, 2015; Marais, Saleh, & Leveson, 2006).

Identifying that worker behavior was a contributing factor in incidents or near misses, however, gives us an opportunity to analyze the contingencies that maintained the deviant behavior. These contingencies (often involving management actions and systems) are the real causes of the human error from a behavioral viewpoint. This kind of analysis can also lead to actionable solutions matched to the causal factors.

What, then, does the term "complacency" add to this analysis? Certainly the authors do not view it as a state that can explain a particular human error, nor as a personal failing. We find complacency more useful as a descriptor for a changing pattern of behavior: a pattern in which formerly safe behaviors begin varying in form, eventually including deviations that elevate the risk of process incidents and/or put frontline workers at elevated risk of injury. It is a term that describes a particular kind of behavioral trend that can occur within the task-related repertoire of frontline workers as well as within the decision-making repertoire of management.

Complacency cannot be a root cause; it is a behavioral phenomenon that itself must be explained by reference to the conditions that produced it. Defining complacency as a behavioral pattern or trend has the advantage of being developmental in nature; the pattern has origins much earlier than the error related to a particular incident. By defining complacency as a behavioral pattern, one can appreciate the originating conditions under which the pattern began to emerge. Typically, when incidents occur a particular deviant behavioral pattern already existed.

As a chronic behavioral trend, complacency has been identified as participating in major process safety disasters. For example, Howlett (2001) described patterns of deteriorating operator training, maintenance, procedural compliance, and supervision across several years that contributed to the catastrophic release of methyl isocyanate gas into the air hovering over a crowded residential community during the 1984 disaster at a Union Carbide chemical plant in Bhopal, India, that killed thousands of civilians in one night.

Complacency in management decision making was evident in the events leading to the 1986 NASA Shuttle Challenger explosion. According to the Report of the Presidential Commission on the Space Shuttle Challenger Accident (1986, Chapter 6):

> Neither Thiokol [the contractor responsible for making the solid rocket boosters] nor NASA expected the rubber O-rings sealing the joints to be touched by hot gases of motor ignition, much less to be partially burned. However, as tests and then flights confirmed damage to the sealing rings, the reaction by both NASA and Thiokol was to increase the amount of damage considered 'acceptable.'

And later in that same chapter, the report stated, "NASA and Thiokol accepted escalating risk apparently because they 'got away with it last time.'" Challenger crew members Jarvis, Onizuka, McAuliffe, McNair, Resnik, Scobee, and Smith were killed when the O-rings of one of the solid rocket motors failed during ascent, leading to the explosion of the adjacent liquid fuel tank and destruction of the orbiter 73 seconds after launch.

There is little doubt that the behavioral trends of complacency participate in process incidents and the injury or death of operators. But how and why do such patterns emerge despite numerous safety regulations, the frightful possibility of injury or death, multiple levels of managers, safety training, and modern industrial sites with advanced control technology? Furthermore, what can be done to combat complacency and increase the probability that operators and managers persist in safe work practices and decision protocols that bolster process safety? In this paper, we will attempt to understand complacency from a behavior analytic perspective with the goal of improving causal analyses and suggesting behavioral intervention strategies that can stop or prevent the pattern from developing.

Behavioral processes contributing to complacency

There are operant and nonoperant behavioral processes that play a role in the origin and continued development of complacency. In this section, we will begin with a discussion of the role of habituation, then address the layers of operant contingencies that contribute to complacency, including discussions of the role of positive and negative reinforcement, variability-inducing processes, various competing contingencies, and rule-governance. These processes will be key to our strategic recommendations for stopping or preventing complacency to be discussed in the second section of the paper. Throughout this analysis, we are seeking to answer one central question: with multiple sources of both positive and negative reinforcement available for strengthening safer work practices and decisions, why do safe behaviors begin to vary and continue to vary until dangerous deviations lead to personal injury, process incidents, or near misses?

Habituation

Habituation is a nonassociative learning process in which the repeated exposure to a stimulus leads to diminished responding (Catania, 2013; Thompson & Spencer, 1966). Repeated encounters with a previously fearful stimulus in a situation where nothing bad happens lessens the fearful responding. For example, exposure-based therapies, such as systematic desensitization, can be intentionally used to reduce or eliminate phobic responses to stimuli such as heights, flying in airplanes, and public speaking (Behnke & Sawyer, 2004).

In the work environment, repeated exposures to stimuli identified as hazardous, which do not result in injury or other type of overt incident, may be expected to produce levels of habituation. Sometimes this habituation may become so complete that the worker or manager no longer experiences a respondent effect (e.g., physiological fear response) to the hazardous stimulus. Cognitively, the perception of potential harm may be diminished despite the fact that the stimulus may continue to be discussed in safety meetings as a hazard to be reckoned with (Geller, 2001).

Many readers have seen the famous photograph taken in 1932 of Manhattan ironworkers casually lunching on an I-beam 800 feet in the air with no fall restraint system present ("Lunch Atop a Skyscraper" by Charles C. Ebbets, 1932). Habituation is exemplified in this photo. While not permissible with today's safety standards, modern construction workers still do their daily jobs experiencing habituation in some form; indeed some level of habituation might be necessary for them to do their jobs in these environments.

The problem is that an individual, manager, or engineer cannot intentionally arrange a moderate level of habituation; one that is enough to maintain fear responses and mitigating behaviors while not being so strong as to interfere with job performance. Habituation appears to be an automatic process running continuously in the background. Thus, habituation may exceed "useful" levels. To the extent that hazards are seen as less and less fearful, the motivation for any protective operant avoidance behaviors is weakened. Thus, habituation may be one process contributing to complacency.

Operant processes in complacency

In contrast to habituation, operant processes involve behaviors coming into contingent association with environmental stimuli preceding (antecedents) and following behavior (consequences) to form units called behavioral contingencies (Catania, 2013; Skinner, 1969). The same behavioral topography can be influenced jointly by more than one contingency. In the work world, most of the human behavior we are interested in participates in multiple contingencies within a context of multiple concurrently available choices (Ludwig, 2002). This is why so many Organizational Behavior Management (OBM) practitioners find behavior analysis tools that examine multiple contingencies within concurrent choice useful for understanding the factors affecting the choice between safe and at-risk behavior (e.g., PIC/NIC Analysis, Agnew & Daniels, 2010; Daniels & Bailey, 2014; Behavior Systems Analysis Questionnaire, Diener, McGee, & Miguel, 2009). It may be worthwhile to look closely at the contribution of different behavioral contingencies to the characteristics of behavioral acquisition, variation, maintenance, and deterioration to understand why trends like complacency may develop.

Positive reinforcement and extinction

For the process industry worker, the typical transition from training to post-training work environments may set the stage for the development of complacency. While a number of personal safety actions involve fairly simple and routine operants such as putting on Personal Protective Equipment (PPE), tasks involved in process safety can be more complex and require more training. Tasks involving large amounts of energy or dangerous chemicals critical to process safety are often highly proceduralized. For example, lock-out/tagout tasks to de-energize electrically powered equipment such as pumps in a chemical plant prior to maintenance work typically require dozens of detailed steps spelled out in written standard operating procedures. These tasks require low variance/high adherence in the execution of engineered process steps designed to control energy or chemical release.

When trainees first learn a complex task that puts them in the presence of substantial hazards, they are often in an environment of dense reinforcement. In training sessions, they find themselves directed with numerous antecedents and supplied with many immediate consequences as they are coached by a trainer, manager, or fellow worker until they demonstrate a degree of mastery to gain certification and the go-ahead to engage in these process safety tasks in the work setting. As procedures are learned, behaviors exhibited in the training setting are full of variance initially, as trainees vary the way they do tasks in big and small ways until the dense antecedents and consequences eventually shape them to do the task the same way every time. Procedural training, then, narrows behavioral variability in procedural adherence from the beginning to whatever endpoint is deemed competent.

How do trained workers lose this low-variance performance and develop patterns of higher variation that we call complacent? Behavior and behavior chains involved in complex tasks can undergo extinction when there is a substantial reduction in reinforcement (Catania, 2013). Basic research has shown that extinction is one circumstance that generates increased variability (Lerman & Iwata, 1996; Mechner, Hyten, Field, & Madden, 1997). After a relatively intensive training period, workers are typically left to engage in the task over and over without the dense training contingencies. Thus, they experience a relative reduction in positive reinforcement and start to exhibit small variations in the way the task is done. Employees begin to glance away from the work, allow for a bit more slack in the line, check off items on the checklist they have not completed (colloquially referred to as pencil whipping, Ludwig, 2014), skip a step in the procedure that appears unnecessary, and many other variations on the safe work procedure.

Eventually, these variant behaviors often find some type of unprogrammed reinforcement (Agnew & Daniels, 2010; Agnew & Snyder, 2008). Perhaps the new reinforcers are available from a small bit of social interaction, escape from boredom, a quicker procedure, or behaviors that are more comfortable or convenient, reducing response cost. These reinforcers may produce small variations at first but additional variants get shaped the same way and the variance gets larger, perhaps unnoticed by the performer or managers due to the incremental shaping. Social processes may help spread the deviation in the population of workers, as coworkers imitate what an experienced worker shows or tells them to do. The phenomenon known as "normalization of deviance" refers to the situation in which deviations are so common that they come to be seen as the norm, treated by workers and even managers as acceptable practice (Vaughan, 1996). What starts as small variation in the behavior of an individual worker can end in acceptance of population-wide deviations among the plant population that pose a chronic, latent risk of injury or process incidents.

Negative reinforcement and avoidance behavior

The behaviors we discussed above aren't solely influenced by increases or decreases in levels of positive reinforcement. Behaviors designed to prevent injury or process incidents are also affected by negative reinforcement contingencies. It is tempting to think that behaviors motivated by avoidance of disastrous outcomes would show low variability and high persistence. Why, then, are supposedly robust avoidance contingencies not sufficient to keep behavior from drifting into dangerous variations? A closer look at avoidance behavior in industrial settings is warranted.

One could argue that avoidance contingencies form a foundation for safety programming. Tasks in industrial settings require myriad safe actions to help employees avoid getting hurt or doing great damage to the physical plant and its surrounding communities. In this light, it is not surprising that the traditional management practices many companies use to boost safety are aversive control techniques: threats of suspension and firing, safety talks highlighting frightening injury potential, and liberal use of actual discipline after injuries or unsafe actions were detected. Thus, companies use avoidance-based management contingencies to generate compliance with safe work procedures. These contingencies are supported in a greater system where governmental regulatory agencies also rely heavily on threats and fines to encourage company leadership to adhere to safety regulations. Workers and managers in process industries operate in a working world rife with negative reinforcement contingencies for safety.

Even with these avoidance contingencies operating, complacency develops. Something is making the avoidance of injury a less powerful reinforcer than laboratory research would suggest. Laboratory studies of avoidance using rats often presented periodic electric shocks (small amperage, high voltage) unless the rat pressed a lever. Sidman (1953) showed that avoidance responding was maintained when it delayed the onset of shock (i.e., the response-shock intervals exceeded shock-shock intervals in his unsignaled avoidance preparation). Other laboratory studies of signaled avoidance suggested that intervals of no electric shock ("safety periods") could reinforce avoidance behaviors (Azrin, Holz, & Hake, 1962). Herrnstein and Hineline (1966) showed that shock-frequency reduction was sufficient to maintain avoidance behavior and suggested frequency reduction of the aversive event as the critical reinforcer in avoidance. In these basic research studies, avoidance behavior appears robust, but they are using parameter settings that do not model some real-world parameters present in industrial settings.

As a consequence for safe behavior, avoiding injury is unlikely to be salient unless it represents a huge change from a baseline of high-frequency injury or process incidents. The laboratory studies provided environments where aversive events happen every few seconds unless the rat made a singular avoidance response. Most modern industrial settings actually have relatively low base rates of injuries and process incidents with months or years separating major aversive events, so the appearance of having avoided injury is the norm, not clearly brought about by specific avoidance behaviors. This would make avoidance a poor candidate as a powerful reinforcer to strengthen safety-related avoidance behaviors in today's industrial settings.

In addition, given the probabilistic nature of injury per each unsafe act, one can appear to "be safe" even when one has just behaved unsafely. Many workers engage in actions that put them at-risk and get away unscathed despite close calls. Similarly, senior leaders may make decisions that increase hazards, yet see no disaster for years. The discriminability of success for avoidance behavior is low in this context: most of the time behaving safely or at-risk results in being safe. For this reason, the absence of injury or incident is not, by itself, likely to strengthen safe actions in most modern industrial environments.

The participation of rule-governed behavior

Rule-governance participates in the acquisition and maintenance of safety-related behaviors and adds a layer of verbal contingencies to take into account. When we observe injury avoidance behaviors of an adult, we see the result of years of exposure to direct-acting contingencies as well as indirect-acting contingencies (Malott, Shimamune, & Malott, 1992; Weatherly & Malott, 2008), with the preponderance being rule-driven, indirect-acting contingencies. No worker has experienced most injuries they are warned about, and no plant manager has experienced every industrial disaster possible at their site. Instead, the majority of their avoidance behaviors have been instructed through training, safety talks, what-if scenario discussions, and the like. Rules of the general form "in this task, if we do these things, we'll be safer; and if we don't do these things we could get hurt this way or have a process incident like this" are taught and self-formulated with reference to multiple specific hazards in any industrial setting.

Rule-following in cases where supervisors, safety professionals, or fellow employees provide instructions on how to avoid injury or discipline on specific tasks would be affected by two different kinds of indirect-acting, rule-based contingencies (Hayes, Zettle, & Rosenfarb, 1989). In the rule-following known as *pliance*, the rule-giver would supply social consequences for following the rule, such as when a supervisor gives feedback on permit compliance to a frontline worker. In the rule-following known as *tracking*, the natural outcomes of following the rule act as reinforcing consequences. For example, in hot work, a person designated to do fire watch (a step required by the permit) discovers a smoldering fire and smothers it; preventing a fire from getting out of control might strengthen the behavior of following the tracking rules prescribed in hot work permits.

Repeated exposures to pliance and tracking consequences strengthen not only the specific instances of rule-following, but also strengthen generalized rule following (Hayes et al., 1989). Supervisors who articulate rules for permit compliance and then supply social consequences for following or violating the rule may find their direct reports more likely to follow other rules articulated. Similarly, if the rules associated with fire watch are strengthened by tracking consequences such as finding a minor fire incident, then other rules specifying tracking outcomes may be strengthened to some degree also.

The avoidance paradox

Avoidance behavior brought about through direct-acting contingencies as well as indirect-acting contingencies rely on at least the occasional occurrence of an aversive event such as an injury or process incident to persist. Studies of direct-acting avoidance contingencies in laboratory preparations showed that if the shock generator were turned off after the acquisition phase, avoidance behavior would diminish (Schnidman, 1968).

In the case of rule-governed avoidance, when the aversive event doesn't occur very often, the credibility of the rule specifying various avoidance behaviors may be diminished; this can weaken rule-following in a generalized fashion. Both forms of rule-following, pliance and tracking, may weaken, though not necessarily to the same degree (something we can leverage in our solutions to behavioral variance in process safety). Hayes et al. (1989) also discussed *augmentals*, rules that function as motivating operations and establish or alter consequence functions. One would also expect that attempts at augmenting to bolster the threat of injury or incident would lose their effectiveness in this context. Augmentals such as sharing details of injuries at other plants or hearing testimonials from previously injured individuals may seem less frightening to workers or managers when the frequency of injuries or process incidents is low.

It is common in industrial settings with good safety records to have workers and managers eschew the need for enhanced safety procedures because "no one has been hurt here in x years." This is the real-world *avoidance paradox*: the more effective avoidance behavior is at preventing injury or incident, the lower the motivation to continue avoiding. The safer we have been, the less safe we become, until injuries or process disasters occur sporadically and re-motivate avoidance.

Competing programmed contingencies

All real-world safety contingencies operate concurrently with other competing reinforcement contingencies (Ludwig, 2002). As we discussed above, sometimes these concurrent contingencies offer unprogrammed reinforcement from informal social interaction, or greater comfort or convenience, that may strengthen behaviors that compete with safe work practices. There are also programmed contingencies that can compete with safety contingencies. The major competing programmed contingencies in industry are those involved in production. Companies don't make safety; they make products or services to sell to the market, and the powerful economics of production generate many strong behavioral contingencies supporting production. The worker is acutely aware of production and schedule demands, the plant manager is held accountable for meeting production goals, and the survival of the company depends on the profits from this production. In this concurrent context, if safety-related avoidance behavior weakens, the ever-present production-related contingencies will strengthen relatively.

The demands of production provide reinforcers for behaviors related to reducing production time and costs. These behaviors may involve taking shortcuts to obtain these savings in time or dollars. Worker actions and management decisions may creep toward the riskier end of the safety spectrum when the avoidance contingencies are relatively weaker. If incidents have been rare, senior operations leaders or financial officers may be reinforced to save money by reducing staff or disapproving requests for capital equipment upgrades, even when safety concerns have been voiced (Ludwig, 2017). These production-centric behaviors, throughout the system, may ebb when there is a process safety incident or near miss, which strengthens the safety-related avoidance behaviors again. All too often, the systemic process of strengthening and weakening safety contingencies repeats in a sine-wave oscillation over many years.

Summary

Many processes and variables combine to weaken safe behavior of workers and managers in industrial settings, inducing and amplifying variance over time to the point where deviations pose a serious risk of process incident or injury--the trend we call complacency. Habituation reduces the aversive characteristics of hazards on the job. Levels of positive reinforcement for safe behavior of both workers and managers are often meager day to day, inducing behavioral variation that can contact other sources of reinforcement, be amplified through shaping, and spread via social processes. Low base rates of incidents and injuries in many modern industrial settings combined with low probability of adverse events per each unsafe act render low discriminability of success for avoidance behaviors. Rules that guide safe actions can be weakened by the rarity of adverse events. Successful safe behaviors reduce the overall motivation for continued avoidance. Finally, production-related contingencies may compete with and override weakened safe behavior of workers and managers. What solutions can our behavior analytic approach offer?

Strategies for preventing or reducing complacency

Simplistic notions of safety based on commonsense understandings of avoidance are not likely to lead to effective action. Informing people of the hazards and hoping that this is sufficient to scare them into permanently behaving safely is not a strategy for promoting successful personal or process safety. Alerting people to the hazards with frightening details can be part of a safety strategy, but if that were sufficient, no one would smoke cigarettes today; it has been over 50 years since blunt warnings about disease and death started appearing on cigarette packs. Other components of a comprehensive safety strategy must be strengthened to make safe behavior more persistent.

Identify and assess behavioral variance

Assessing behavioral variance at the frontline employee level

If behavioral variance is a factor in process safety, we must, as a first step, have a method to assess behavioral variance before any interventions are implemented. These assessments may include (a) identifying specific work processes that have high potential for a process incident, (b) conducting observations of the tasks following a behavioral checklist derived from operating procedures, and (c) comparing the existing levels of behavioral variability to desired or tolerable levels. Video of actual work tasks can be used to identify acceptable and unacceptable variance with focus groups of employees who perform the tasks (note that a no-retribution ground rule facilitates honest discussion).

Identifying high-risk work processes is already part of most process safety efforts, so that would not be new. Viewing *behavioral* variation as a potential hazard itself and examining levels of behavioral variance would be new to some, but it could be seen as an extension of current practices of viewing other forms of variation as a potential problem. For example, it is common for hazard analyses in heavy industry to include a *safe operating limits* analysis to understand the boundary between a level of acceptable variation considered "normal operations," a buffer zone of increasing variance called the "margin of safety," and a "failure zone" where variance is so large that injury or process incident is likely (see Howlett, 2001, pp. 57–69 for a discussion of boundaries of operation in industrial settings). Safe operating limits are often mapped out for key processes in manufacturing or chemical processing where, for example, control room operators need to know which temperatures and pressures are within a normal range of operations and which are outside that zone and require some intervention. The logic of safe operating limits could be applied to behavioral variation as well.

Some process-related work tasks would have smaller *safe behavior operating limits* than others. In a process industry such as petrochemical refining, "breaking lines" means opening pipes that carry product in order to inspect them, replace a stuck valve, and other maintenance actions. Adding a "blind" or "blank" flange to temporarily close the open pipe is a critical step to prevent product from escaping should an upstream valve open during maintenance operations. If this step were to be skipped or if the flange were bolted unevenly, the workers would be at risk of being injured by toxic product and there would be a risk to the wider population because of the loss of primary containment (LOPC) of a toxic or flammable liquid or gas. Therefore, even small variations in installing the flange, such as in bolt sequencing and torqueing, could be outside of safe limits. Employee teams could identify what steps in the breaking line process have the most risk of process safety incident. Upon viewing videos of this process they may identify that many peers install the blank flanges differently (i.e., variance). These same peers might then be able to design the safe behavior operating limits by discriminating critical behavioral steps (e.g., lining up gasket, torqueing patterns). These safe behavior operating limits then can be used in training, job instructions, and on-the-spot labels.

Where behavior-based safety (BBS) systems are in place, many people observe worker(s) doing risky jobs, record whether the behavior was safe or at-risk, and have feedback discussions with the observed worker(s) afterward. BBS systems would already have the data streams to examine frontline worker variance. Process safety guidelines from the Center for Chemical Process Safety (CCPS) have advocated that conformance to procedures be audited through field observations of workers performing critical tasks (CCPS, 2007), so examinations of behavioral variability at the frontline worker level are not a new idea in process safety. Thus, we strongly recommend that process safety behaviors such as inspecting critical equipment should be explicitly added to BBS observation cards.

Assessing behavioral variance at the management level

Variation in supervisory practices can impact process safety risk as well. Critical behavioral antecedents such as job safety briefings can be audited to assess if supervisors cover all the key points to prepare workers to control process safety issues effectively. Minor variations in style, such as the use of humor or examples, may be considered within safe behavior operating limits. If, however, some supervisors fail to discuss key process safety issues facing workers some of the time, this may be considered a level of variation outside of safe behavior operating limits and should be addressed by management.

Table 1 shows some sample management behaviors and decisions that could drift outside of safe behavior operating limits sporadically, or as part of a deteriorating trend, that could lead to process safety events. Thorough study of major process safety disasters like those we mentioned at the beginning of this article often reveals these and other management actions as contributing or root causes (c.f., Howlett, 2001).

Some system would need to be developed for examining drift in critical management behaviors and decisions affecting process safety. Corporate safety officers or outside consultants could review management decisions

Table 1. Sample Complacent Management Actions and Their Safety Impact.

Management actions	Adverse safety impact
Letting Process Safety Management (PSM) program deteriorate	Management of change (MOC) process slows down; old less safe procedures continue
Reducing staffing levels or expertise	Mistakes by fatigued or inexperienced workers increase
Increasing production pressures without additional capacity	Safe work practices short cut to meet production/ schedule demands
Spending increasing time in the office, less time out in the process areas; dwindling management oversight of safety	Managers less aware of, or don't take actions to deal with, existing safety threats
Pressuring maintenance department to just "patch it and keep it running" versus proper overhaul or replacement	Results in poorer mechanical integrity of the plant; risk of mechanical failure increases, risk of human error interacting with poor equipment increases
Using poor behavior management practices	Failure to change at-risk behaviors, incentivize the wrong actions, drive up resentment of management so that rules are less likely to be followed

through discussion or documentation periodically. For example, it would be relatively easy to detect when management reduce staffing to levels that could be a threat to process safety in that setting. Other subtle management behaviors might be revealed through interviews and survey responses of their direct reports regarding actions such as increasing production pressures to the point of overwhelming safe practices.

Behavioral interventions to reduce complacency

Several interventions could prevent or reduce behavioral variance when it is outside of safe behavior operating limits. There are more precedents in OBM for addressing the behavioral variance of frontline workers than precedents for changing variation or drift in management behavior, although both strategies are needed.

Variation reduction training

Training design could be improved so that more emphasis is put on reducing behavioral variation. Trainees should learn to perform tasks within safe levels of variation, with a focus on the boundaries of safe variation for each task. That is often explained to trainees in some form, but it is less common for training to include sufficient practice that trainees actually learn to recognize and mitigate behavioral variation. Trainees can be led to experience variation in their behavior and be taught to recognize when it happens to themselves and coworkers. Brief but complicated tasks can be set up in training with outside stimuli, such as other work demands or coworkers wanting to socialize, added to induce behavioral variation. Fellow trainees could formally observe behavioral variation in the task to gain experience recognizing, analyzing, and mitigating variation.

Fluency-based training

Fluency-based training models include repetition at real world speed with a focus on accuracy, which address variability along two dimensions. Fluency-based training may help reduce unwanted variation and strengthen persistence well beyond the training setting (see Binder, 1996). Athletic training and military training (e.g., disassembling and reassembling weapons) are domains where fluency training is more common than the average workplace at present. Fluency-based training could be used, for example, to train workers to spot workplace hazards and identify the best hazard mitigation steps with accuracy and speed, lowering the effort it takes to do hazard assessments on the job and thus increasing the probability of that behavior in the field. Because safety-focused pretask inspections are such a common element in personal safety (e.g., pretrip vehicle inspections) and process safety (e.g., scanning the work environment for hazards before opening a valve), fluency-based training could be a useful tool for trainers in heavy industry, where it has been seldom used, striving to reduce unwanted behavioral variation. Fluency-based training exercises also have one more advantage over more conventional training exercises according to Binder and Sweeney (2002): they take less total training time and can be delivered in fun game-like formats.

Enhancing reinforcement for safe behaviors

Training, by itself, is no silver bullet (Hyten, 2015). To combat extinction in the post-training environment, behavior would have to be supported through follow-up reinforcement mechanisms (Wick, Pollock, Jefferson, & Flanagan, 2006). Decades ago, behavioral psychologists realized that both natural and planned levels of positive reinforcement were low for safety-related behaviors in most industrial settings, and that the unwanted side effects of reliance on aversive safety management practices were crippling advances in safety. Strategies utilizing positive reinforcement to directly and indirectly strengthen safer practices are available and currently in use.

Behavioral approaches to safety, including systems such as BBS, were developed to promote deliberate use of positive reinforcement for specific safe behaviors, and these approaches have become widespread (Geller, 2001; McSween, 2003). Where such systems are in place, positive reinforcement from peer feedback, achieving group goals, and improvements to the workplace combines with the negative reinforcement contingencies supporting avoidance behaviors. The addition of positive reinforcement can counter some of the negative side effects of avoidance contingencies based on negative reinforcement, and BBS systems in particular have often produced high levels of enthusiasm and loyalty among workforce participants (Geller, 2001).

BBS systems are designed to provide relatively immediate, though intermittent, positive reinforcement to frontline workers for observed safe acts and constructive feedback without formal penalties for observed at-risk actions. This level of stepped-up feedback can help keep behavioral variation in check for the behaviors being observed, reducing the chance of dangerous deviations developing and becoming widespread. Therefore, process-risk task-specific observation cards can be developed based on safe behavior operating limits. Workgroups can then use these specialized cards to observe and coach each other until all behaviors are within operating levels. To accomplish this, peer observers need to attend closely to potentially subtle variations in behavioral form or task sequence to determine whether the actions they are observing are within safe behavior operating limits or drifting outside these limits. Post-observation feedback discussions between observers and workers include information about behavioral variance. Detailed emphasis on behavioral variance may be a new feature in some

BBS systems, especially ones where observers have become used to just giving binary feedback that observed actions were safe or not.

BBS systems can be more intentionally adapted to promote process-safety related behaviors that typically are not included in BBS observation cards. Most workers in chemical and high-energy industries work in and among the very equipment and controls designed to manage process safety. They are therefore in the position to scan their environment for conditions that seem out of place (e.g., corrosion on piping) or for behaviors that may contribute to process safety controls (e.g., whether workers follow procedural steps). BBS observations thus adapted could enhance process safety by promoting more frequent conversations, data collection, and interventions around physical conditions that might otherwise wait for scheduled inspections, and at-risk behaviors, which are often addressed less directly in standard Process Safety Management (PSM).

Another advantage of safety approaches that include feedback and reinforcement for safe behaviors in the field, not all of which fall under the heading of BBS, is that the emphasis on behavior instead of outcomes can help deal with the discriminability problem we discussed in the previous section. Relying on an outcome such as the avoidance of injury or "being safe" as the principal consequence to reinforce safe actions is not likely to be a powerful contingency in environments where adverse safety events are relatively rare anyway. Shifting the emphasis from avoiding injury or incident to whether the behaviors or tasks were executed properly is a much more easily discriminable criterion of success for both worker and observer.

Reducing reinforcement for unwanted behavioral variation

Deviation from safe practices may be maintained by reinforcement such as saving time or effort. If the motivation for taking a risky shortcut is the response cost of a cumbersome process or procedure, an effort to streamline it may reduce the time or effort savings, minimizing the gain for taking a risky shortcut. Equipment and tools play a role in task time, so they should be examined as part of a streamlining effort. If the process or procedure is already as lean as it can be, performer fluency in the procedure may be an avenue to reduce time or effort savings for at-risk behavior.

The petrochemical industry is well-known for its extensive operating procedures and use of checklists. In complicated process safety tasks, such as those that involve preventive maintenance, the checklists can be dozens of items long. The second author has witnessed checklists over 100 items long for a high-variance yet critical task. When asked to see completed copies of this checklist, systemic pencil whipping (Ludwig, 2014) was evident (the checklists had been photocopied with the date changed). Pencil whipping was negatively reinforced by avoiding extra time and effort and this was associated with extra deviance in preventive maintenance required for process safety success. Deviation was substantially reduced when the checklist was downsized to only the behaviors that were most variant and most time-critical. When the response cost was greatly reduced, so was the negative reinforcement for the deviations.

Strengthen contingencies that support rule-following

Following safety rules and safe operating procedures, assuming they are up-to-date, is integral to process safety. Rule-following may have pliance and tracking contingencies simultaneously influencing it; strengthening these elements would improve the likelihood of consistent rule-following and reduce variation outside of safe behavior operating limits. Because it relies on social consequences rather than safety outcomes, pliance has the potential to better withstand the demotivating effects of the avoidance paradox discussed earlier.

To enhance pliance, supervisors and managers can be more diligent in providing reinforcement to each other and their direct reports for following important safety rules. That could include, for example, manager-to-manager feedback for meaningful contributions to PSM rules, such as recognizing a manager for finding a process to expedite implementation of management of change (MOC) committee decisions. Supervisors can regularly provide positive or constructive feedback to frontline workers for adhering to critical operating procedures (e.g., "we had a great day today because we executed each of the procedure steps perfectly"). Peers could also provide feedback to their coworkers on procedural compliance (e.g., "we did that job the right way because we stuck to the rules"). Any feedback that recognizes forms of rule-following would bolster pliance. Even questioning unclear procedures in order to yield better rules should be reinforced. Simply stepping up punishment for deviating from procedures would not be advisable without enhancing positive reinforcement for complying.

Tracking contingencies rely on natural outcomes to reinforce rule-following. If the outcomes specified in tracks are the avoidance of injury or incident, then the tracking contingencies will be more vulnerable than pliance contingencies to being weakened in environments where adverse events are rare because of habituation to the threat of hazards. Tracking contingencies also can suffer from the discriminability problem we discussed earlier whereby both safe and at-risk actions most often produce no adverse event, so the rule appears to be of questionable accuracy or utility. Sharing and analyzing company or industry injury reports, near-miss reports, what-if discussions, and after-action debriefs may make the dangers and personal benefits of following safety rules clearer. Even if this produces only a modest reduction in habituation and helps to augment the tracking contingencies to some degree, it may still be worthwhile to do. Many companies employ near miss reporting systems at some level now, but the quality of these efforts could be increased (Agnew & Uhl, 2015).

Celebrating successful safety outcomes is common in many companies. For these celebrations to have any behavior-strengthening effect, because they are infrequent and delayed from daily safety actions, it is important to enhance the rule-governed processes that exist in them. All too often, celebrations spend too much time talking about the outcome (e.g., no lost-time injuries, or no LOPCs this quarter) without discussing the behaviors that produced those outcomes. We should recognize celebrations as indirect-acting analogs to reinforcement that rely on verbal processes to have any effect on safety-related behaviors. If company celebrations spent more discussing what people did that produced the excellent outcome, rules relating behavior to safety outcomes could be formulated or bolstered and rule-following thereby strengthened.

Reinforcement of interlocking management behaviors

Historically, much of OBM practice in general (safety in particular) has focused on behavioral interventions targeted at frontline workers (Hyten, 2002, 2009). To deal comprehensively with complacency, systems that target the safety-related decision making of managers must be included. Management hierarchies are already designed to enable feedback about decisions from within and across management levels, though it is more common for feedback to be top-down. Even top-down feedback would be helpful if what is discussed goes beyond production-centric topics to how well safety and production are integrated and affected by particular decisions. There is a history evident in industrial disasters of production contingencies overriding safety considerations (cf. NASA Shuttle Challenger incident), so robust systems to monitor and change drift toward unsafe decisions of managers and executives have yet to be demonstrated to the degree that frontline-focused BBS systems have been vetted.

Companies should weigh safety management together with productivity in selecting and evaluating managers. If a manager increased safety risk but was evaluated highly and promoted principally because of increased production, then these reinforcement contingencies will be clear to that manager and every other manager in the company. If we are striving to address complacency at all levels in companies, management evaluation and promotion criteria must not reinforce dangerous deviations in safety-related decision making.

Certainly regulatory agencies such as the Occupational Safety and Health Administration (OSHA) in the United States create metacontingencies (Malott & Glenn, 2006) for managerial actions related to safety and increasingly to process safety. Artifacts of manager actions that violate these rules found during incidents, inspections, or reporting can result in large fines, operating limitations, further scrutiny, and the deterioration of company reputation. These company-level contingencies impact individual manager contingencies as they may see themselves disciplined, their bonuses negated, or their job security expired. The reality is, however, that heavy penalties for patterns of safety violations have been relatively rare, so even this negative reinforcement-based approach to changing management behavior is weaker than it sounds on paper.

One avenue to be explored is the role of groups who advise or oversee senior management decisions, such as board of directors. Senior executives must report their courses of actions to these boards periodically and they can put considerable pressure on executives to improve the financial success of the company. Because these boards essentially set behavioral contingencies for senior management, and thus the rest of the company, they have the opportunity to shape better integration of safety and production in their discussions with executives. Given that boards of directors focus considerable time on financial results and company survival in the marketplace, a starting point for such work might be educating board members not only in the massive direct and indirect costs to the company for process disasters (which is presumably being done), but also how the interlocking behavior of senior leaders, managers, supervisors, and frontline employees contributes to these disasters. Helping board members to see how the contingencies of production and cost control that they foster can influence the proper integration of safety with production might alter the nature of their discussions with senior leadership in ways that better promote process safety.

Conclusions

Complacency is best understood from a behavioral perspective as trends in the levels and nature of behavioral variation for workers and managers. When such variation exceeds safe behavior operating limits, safety is compromised and the risk of process incident and/or injury is increased. We have discussed many behavioral processes that contribute to such dangerous behavioral variation at the level of frontline workers, supervisors, managers, and even senior leaders. Our analysis led us to suggest multiple strategies to counter the drift toward complacency. Some of the solutions are already being used in some form in companies today; others have been rarely used to date. Clearly, there is a need to study these processes and solutions empirically in the field to determine the relative contribution of each of the behavioral processes to the phenomenon, and to determine the relative effectiveness of our suggested solution strategies.

Studying the origins of behavioral variation has been of interest to behavioral researchers in both basic and applied areas for many years (see Antonitis, 1951; Lerman & Iwata, 1996; Ludwig, 2002). Extending this interest to examine methods for controlling unwanted or dangerous behavioral variation in work settings deserves more of our efforts. The U.S. Navy's "Blue Angels" elite pilots meet after every airshow to review how each element in the airshow was completed, for both entertainment quality and their own safety as pilots of high-performance jets that fly within feet of each other (Kelly, 2005). Tiny variations in flying their aircraft have led to over 20 fatal crashes, so they reduce variance by focusing their feedback on technique in these meetings. Astronaut Jim Wetherbee, the most experienced space shuttle commander in NASA history, reported that he kept a log of his own personal errors in an attempt to find trends and reduce variance (Wetherbee, 2015). Attempts to develop practical control methods for behavioral variance obviously are underway in many high-hazard occupations. There is much for the OBM community to do in the coming years to gather data in various forms that could help those who work in process industries find better solutions to an issue that has plagued their considerable safety efforts over decades.

References

Agnew, J., & Daniels, A. (2010). *Safe by accident: Leadership practices that build a sustainable safety culture.* Atlanta, GA: Performance Management Publications.

Agnew, J., & Snyder, G. (2008). *Removing obstacles to safety: A behavior-based approach.* Atlanta, GA: Performance Management Publications.

Agnew, J., & Uhl, D. (2015). *Near-miss reporting: Applying behavioral science to optimize safety culture.* Paper presented at the annual ASSE conference, Dallas, TX.

American Chemistry Council. (2017). *Definition of process safety* Retrieved March 29, 2016, from https://responsiblecare.americanchemistry.com/Performance-Results/Safety#process

Antonitis, J. J. (1951). Variability of response in the white rat during conditioning, extinction, and reconditioning. *Journal of Experimental Psychology, 42,* 273–281. doi:10.1037/h0060407

Azrin, N. H., Holz, W. C., & Hake, D. (1962, June). Intermittent reinforcement by removal of a conditioned aversive stimulus. *Science, 136,* 781–782. doi:10.1126/science.136.3518.781

Behnke, R. R., & Sawyer, C. R. (2004). Public speaking anxiety as a function of sensitization and habituation processes. *Communication Education, 53*(2), 164–173. doi:10.1080/03634520410001682429

Binder, C. (1996). Behavioral fluency: Evolution of a new paradigm. *The Behavior Analyst, 19,* 163–197.

Binder, C., & Sweeney, L. (2002). Building fluent performance in a customer call center. *Performance Improvement, 41,* 29–37. doi:10.1002/(ISSN)1930-8272

Carroll, J. S., Rudolph, J. W., & Hatakenaka, S. (2002, April). *Organizational learning from experience in high-hazard industries: Problem investigations as off-line reflective practice* (MIT Sloan Working Paper No. 4359-02). Retrieved from http://ssrn.com/abstract=305718

Catania, A. C. (2013). *Learning* (5th ed.). Cornwall-on-Hudson, NY: Sloan Publishing.

Center for Chemical Process Safety (2007). *Guidelines for risk-based process safety.* Hoboken, NJ: John Wiley.

Daniels, A., & Bailey, J. (2014). *Performance management: Changing behavior that drives organizational effectiveness* (5th ed.). Atlanta, GA: Performance Management Publications.

Diener, L. H., McGee, H. M., & Miguel, C. (2009). An integrated approach to conducting a behavioral systems analysis. *Journal of Organizational Behavior Management, 29*, 108–135. doi:10.1080/01608060902874534

Geller, E. S. (2001). *The psychology of safety handbook*. Boca Raton, FL: CRC Press.

Hayes, S., Zettle, R., & Rosenfarb, I. (1989). Rule-following. In S. C. Hayes (Ed.), *Rule-governed behavior: Cognition, contingencies, and instructional control* (pp. 191–220). New York, NY: Plenum.

Herrnstein, R. J., & Hineline, P. N. (1966). Negative reinforcement as shock-frequency reduction. *Journal of the Experimental Analysis of Behavior, 9*, 421–430. doi:10.1901/jeab.1966.9-421

Howlett, H. C. (2001). *The industrial operator's handbook: A systematic approach to industrial operations* (2nd ed.). Pocatello, ID: Techstar.

Hyten, C. (2002). On the identity crisis in OBM. *The Behavior Analyst, 3*(3), 301–310. doi:10.1037/h0099982

Hyten, C. (2009). Strengthening the focus on business results: The need for systems approaches in organizational behavior management. *Journal of Organizational Behavior Management, 29*(2), 87–107. doi:10.1080/01608060902874526

Hyten, C. (2015, October). *Is re-training your kneejerk reaction to incidents? Avoiding the training trap*. Paper presented at the annual conference Behavioral Safety Now, Reno, NV.

Kelly, B. J., Director. (2005). *Blue angels: A year in the life*. Arlington, VA: Henninger Productions.

Lerman, D., & Iwata, B. (1996). Developing a technology for the use of operant extinction in clinical settings: An examination of basic and applied research. *Journal of Applied Behavior Analysis, 29*, 345–382. doi:10.1901/jaba.1996.29-345

Ludwig, T. D. (2002). On the necessity of structure in an arbitrary world: Using concurrent schedules of reinforcement to describe response generalization. *Journal of Organizational Behavior Management, 21*(4), 13–38. doi:10.1300/J075v21n04_03

Ludwig, T. D. (2014). The anatomy of pencil whipping. *Professional Safety, 59*, 47–50.

Ludwig, T. D. (2017). Behavior systems: Behaviors interlock in complex process safety meta-contingencies. *Journal of Organizational Behavior Analysis, 37*(3-4), 224–239.

Malott, M. E., & Glenn, S. S. (2006). Targets of intervention in cultural and behavioral change. *Behavior and Social Issues, 15*, 31–56. doi:10.5210/bsi.v15i1.344

Malott, R. W., Shimamune, S., & Malott, M. E. (1992). Rule-governed behavior and organizational behavior management: An analysis of interventions. *Journal of Organizational Behavior Management, 12*(2), 103–116. doi:10.1300/J075v12n02_09

Marais, K., Saleh, J., & Leveson, N. (2006). Archetypes for organizational safety. *Safety Science, 44*(7), 565–582. doi:10.1016/j.ssci.2005.12.004

McSween, T. E. (2003). *Value-based safety process: Improving your safety culture with behavior-based safety*. Hoboken, NJ: Wiley-Interscience.

Mechner, F., Hyten, C., Field, D., & Madden, G. (1997). Using revealed operants to study the structure and properties of human operant behavior. *The Psychological Record, 47*, 45–68.

Occupational Safety & Health Administration. (2000). *Process safety management*. Retrieved October 13, 2015, from https://www.osha.gov/Publications/osha3132.html

Report of the Presidential Commission on the Space Shuttle Challenger Accident (1986). Retrieved June 16, 2016, from http://history.nasa.gov/rogersrep/v1ch6.htm

Schnidman, S. R. (1968). Extinction of Sidman avoidance behavior. *Journal of the Experimental Analysis of Behavior, 11*, 153–156. doi:10.1901/jeab.1968.11-153

Sidman, M. (1953). Avoidance conditioning with brief shock and no exteroceptive warning signal. *Science, 18*, 157–158. doi:10.1126/science.118.3058.157

Skinner, B. F. (1969). *Contingencies of reinforcement: A theoretical analysis.* New York, NY: Appleton-Century-Crofts.

Thompson, R. F., & Spencer, W. A. (1966). Habituation: A model phenomenon for the study of neuronal substrates of behavior. *Psychological Review, 73,* 16–43. doi:10.1037/h0022681

Vaughan, D. (1996). *The Challenger launch decision: Risky technology, culture, and deviance at NASA.* Chicago, IL: University of Chicago Press.

Weatherly, N., & Malott, R. W. (2008). An analysis of organizational behavior management research in terms of the three-contingency model of performance management. *Journal of Organizational Behavior Management, 28,* 260–285. doi:10.1080/01608060802454643

Wetherbee, J. (2015, October). *Controlling risk in a dangerous world.* Paper presented at the annual conference Behavioral Safety Now, Reno, NV.

Wick, C., Pollock, R., Jefferson, A., & Flanagan, R. (2006). *The six disciplines of breakthrough learning: How to turn training and development into business results.* San Francisco, CA: John Wiley & Sons.

Behavioral Perspectives on Variability in Human Behavior as Part of Process Safety

Angela R. Lebbon and Sigurdur O. Sigurdsson

ABSTRACT
Process safety involves worker decisions at various points in an extended process, and much remains unknown regarding sources of variability in worker behavior at these decision points. This paper seeks to explain why some workers may be deviating from sanctioned policies and procedures. Risky choice is analyzed through discussion of positive and negative reinforcement, habituation in terms of respondent and operant behavior, risk discounting, and consequence dimensions that include a review of prospect theory, heuristics, and behavioral decision theory. Recommendations are made for improving our understanding of sources of variability in process safety by conducting systematic research on the perspectives reviewed.

Process safety is a relatively new area of interest to behavior analysts, and industries are seeking our help to explain the contingencies and other variables involved in workers failing to adhere to process safety standards established by their companies and government safety agencies (Bogard, Ludwig, Staats, & Kretschmer, 2015; OSHA, 2000). Behavior analysts can explain why workers deviate from process safety standards, and offer methods for studying this area to aid in minimizing and preventing these deviations in the future.

One salient example of such deviations comes from the investigation of the factors contributing to the Deepwater Horizon oilrig spill in 2010, in which 11 people lost their lives. In a report released in 2011 by the Deepwater Horizon Study Group of the Center for Catastrophic Risk Management, it was revealed that a history of at-risk safety choices by workers on the oilrig had contributed to the spill:

> *According to testimony provided at joint investigation hearings, various alarms and critical safety systems failed to operate as intended. It was testified that several of the Deepwater Horizon's fire and gas detectors were not functioning* or had been inhibited prior to the explosion to avoid waking crewmembers in the middle of the night due to false alarms. *This implied that the sensors were able to detect*

hazards and forward the information to the computer. However, the computer would not have automatically triggered the alarm upon detecting a hazard, requiring manual activation for any further response. The failure of the alarms along with other critical safety systems potentially compromised the time available to the crew to evacuate the rig. (p. 47; italics added)

This testimony illustrates how process safety involves a series of worker choices in various work environments. This paper presents different analyses for at-risk choices that contribute to process safety issues, such as behavior maintained by consequences, habituation to an eliciting stimulus, risk discounting, and consequence dimensions (see Table 1 for a summary of these perspectives). In addition to behavior-analytic conceptualizations of worker choice in process safety, we discuss prospect theory, behavioral decision theory, heuristics, and other research conducted outside the field of behavior analysis that has attempted to explain decision-making processes underpinning choices. Finally, we present a contingency diagram (see Figure 1) for one process safety example to illustrate the complex relationship between stimuli, response choices, and consequences that likely follow each choice. Discussion will center on opportunities for researchers to study these variables affecting safety-related choices in process safety.

It should be noted that the behavioral viewpoints expressed here are not intended to focus on workers as the cause of adverse events. Instead, we suggest different viewpoints to potentially explain how worker behavior comes under the control of environmental stimuli and conditions in a manner consistent with a variety of behavioral theories. Those stimuli and

Table 1. Different Perspectives of Variability in Safety Behavior as Part of the Safety Process

Perspective	Description	Example
Behavior maintained by its consequences	A. Positive reinforcement – Saving time – Saving effort	Using unsanctioned tool nearest you, do not have to inspect equipment
	B. Negative reinforcement – Escaping aversive sound – Escaping aversive light	Press button to override warning sounds / lights
	C. Negative reinforcement – Avoiding production loss	Process/work continues on schedule/rate to meet production goals
Habituation to an eliciting stimulus	Rapid/frequent presentation of an aversive stimulus	Stops eliciting the startled response
	Gradual, repeated presentation of aversive stimulus over time	Do not have to investigate cause of alarm, no exploratory activity
Risk discounting	Certain environmental variables affect safety choices in a predictable manner	As effort of safety choice increases, risky choice increases
Consequence-dimensions	Time and probability of consequence influence safety choice	Consequences that are immediate and certain exert greater influence over safety behavior
Verbal behavior	Rules, standard operating procedures, work instructions, directives	Verbal behavior occasions responses when stimuli, that have been previously reinforced, are presented

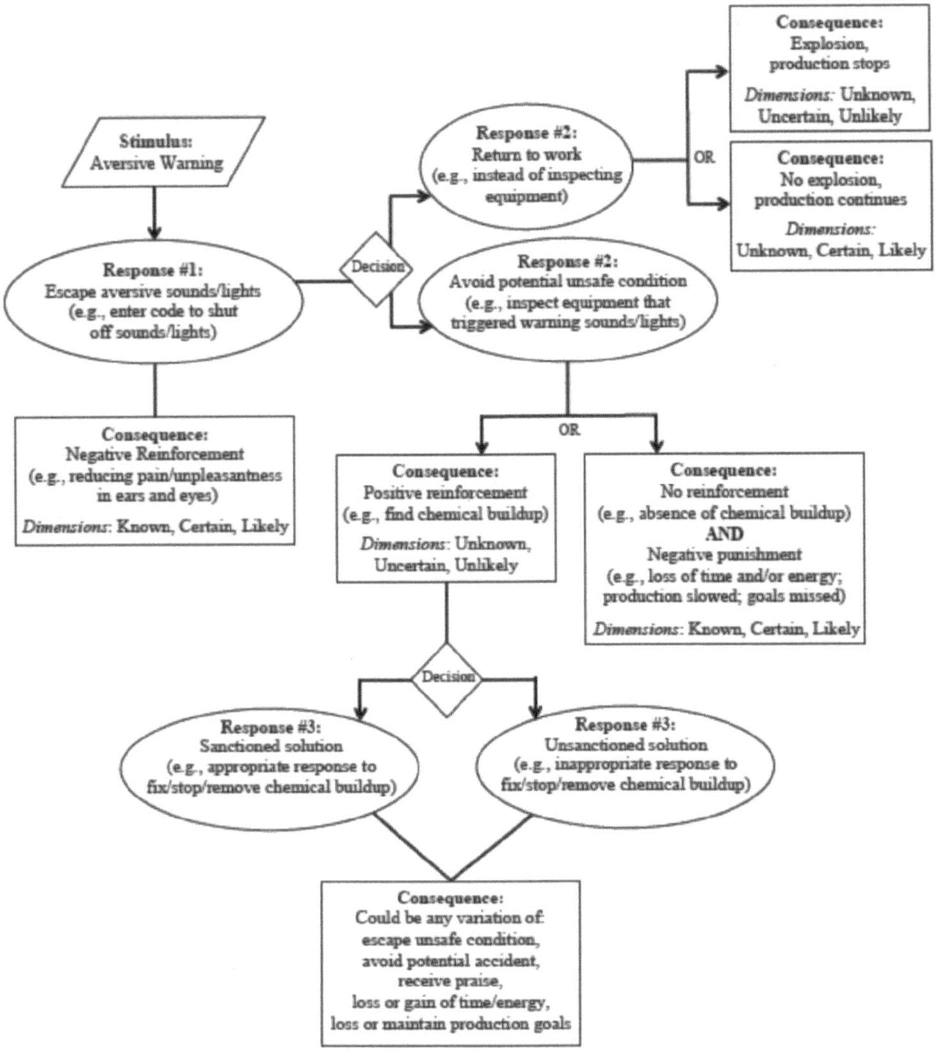

Figure 1. Contingency diagram. *Note.* Before each response, verbal behavior may be emitted that occasions a particular response based on one's reinforcement history and progression through the sequence of situational awareness (see Killingsworth et al., 2016 for in-depth analysis).

conditions may include production pressure from management, faulty warning mechanisms, and fatigue, to name a few.

Behavior maintained by consequences

Positive reinforcement

Little behavioral research to date examine why workers deviate from process safety standards, even though such deviations can be fatal as evident in the

Deepwater Horizon oil spill. Bogard et al. (2015), note that workers at some stage become "complacent," either failing to appropriately respond to warning stimuli (e.g., auditory and/or visual warnings such as loud beeping sounds and flashing lights), or failing to fix the safety issue with the appropriate solution (e.g., using a nonsanctioned, self-devised tool to fix equipment, instead of the recommended tool). A behavior analyst would examine these scenarios and seek to explain the worker's behavior as possibly being under the control of contingencies present in the environment, and those contingencies involve workers saving time and energy by using a quicker and/or easier solution.

Saving time and energy is likely to reinforce workers' nonsanctioned solutions. When workers engage in "short-cutting" behavior (e.g., using the nonsanctioned tool nearest to them instead of a sanctioned tool that requires leaving the work area and locating the tool) and the potential negative consequence does not occur (e.g., no malfunction, injury, etc.), that slight deviation in behavior may be reinforced in the form of saving time, saving effort, and/or meeting production demands. Additionally, each minor deviation does not necessarily result in an immediate, negative consequence(s), but over time, each slight deviation has a cumulative effect that can result in a negative consequence when a threshold is exceeded that is needed to maintain safety.

An employee may override a warning signal to continue the flow of a liquid chemical into a chamber in order to maintain production demands. The chemical reaction occurring in the chamber has known margins of error and increasing the volume past the safety threshold results in an increase in the temperature and pressure due to an increased chemical reaction rate. The slight increase in volume (and subsequent temperature) by overriding one or two warning signals does not result in the negative consequence, but the cumulative effect of repeatedly overriding numerous warning signals results in a buildup of gases that breaches a safety threshold, thus causing an explosion the chamber cannot withstand.

An employee's deviation may become greater over time as the potential negative consequence still does not materialize.

An oil refinery worker climbs a ladder to inspect a malfunctioning hydraulic machine (i.e., leaking fluid and not engaging) during the removal of a bundle hoisted 12 feet from the ground. The worker stands on the third rung from the top with both feet planted on the rung, his body is vertical to the ground as he reaches up to inspect the hydraulic, holding on to the bundle with both hands. As the worker is not in a position to fully inspect the hydraulic he climbs one step up, both feet planted on the ladder but this time his body is in a horizontal position as he leans right and left to get a better view, but with only one hand on the bundle to stabilize his weight. The worker spots the problem so he removes his right foot off the ladder and places it on the bundle for a leveraged view, with only one hand holding onto the bundle. Over the course of this sequence of behaviors, the worker's deviation increases in magnitude. The worker is now engaging in a number of unsafe ladder behaviors that increase his chances of falling, including increased heights and position instability, which requires a harness to hook to the bundle for fall protection once one foot was removed from the ladder and one hand from the bundle.

This example illustrates that workers may be deviating from safety protocols through local contingencies in the environment that they are not even immediately "aware" of, and the deviation becomes harmful only after a cumulative effect or when a certain magnitude of deviation is reached.

Negative reinforcement

Worker "short-cut" behavior may indeed be positively reinforced. However, workers are also likely to experience concurrent negative reinforcement when they behave to escape aversive warning stimuli (e.g., loud beeping sounds and flashing red lights) often designed into process safety management. Escaping the aversive warning stimuli (stopping the beeping sounds and flashing lights) is the behavior that is reinforced first.

If a machine's computer systems detect a potential unsafe condition (e.g., buildup of chemicals or malfunctioning alarm), warnings are emitted that serve as antecedents signaling danger. This may prompt a response to escape those aversive stimuli, in addition to escaping the potential unsafe condition itself.

In this scenario, it is likely that the worker is first responding to the computer's warning systems in a way that functions to escape the aversive sound/light conditions by turning off the warning alarm. This would likely happen before any behaviors related to inspecting equipment and fixing the chemical buildup (see Figure 1 for illustration). Therefore, a sequence of responses is required: the first response escapes warning sounds and lights, the second response attends to the potential faulty equipment, and the third response fixes the discovered faulty equipment. In this sequence of responses initiated by the aversive warning stimuli, the third response of seeking (and finding) a quicker/easier solution to fix the mechanical problem is the last behavior to be reinforced in the sequence. The negative reinforcement that shaped escaping the signals potentially interferes with eliminating the second threat (the potential unsafe condition).

Moreover, the alarm is an antecedent to prompt entering the code to escape the warnings, but there are no antecedents to prompt inspecting the equipment. It is a chaining procedure, per se, but if the next response in the sequence is not reinforcing, then another antecedent needs to be put in place after the first response. For example, a verbal prompt by the operator's supervisor instructing the operator to inspect the equipment after responding to the computer's warning signals may serve that purpose. Alternative methods or chains to ensure actual safety checks before a system is re-energized need not necessarily involve a supervisor, but the check would have to be designed in the process to avoid errors of this kind.

Warning stimuli function to signal the delivery of other, more aversive stimuli. Throughout our lives, rules[1] have been created around the purpose of these warning stimuli.

Children are trained in school to recognize loud beep sounds in coordination with rapid, flashing lights as warning stimuli for a fire in the building (i.e., another, more aversive stimulus). They also learned through practice that the appropriate response to such warning stimuli is evacuating the building. They were told that these behaviors save their lives when burning and death are imminent but it is highly unlikely they ever actually avoided the natural consequences.

One quickly learns that these warning stimuli in coordination with evacuations are merely practice, and the negative consequence of burning in a fire does not occur every time the warning stimuli appear. In fact, with the sight and smell of fire not present when these warning stimuli are presented again in the future, verbal behavior among the individuals exposed to these warning stimuli may suggest a lack of discrimination, "is it a real fire or a fire drill?".

The history of consequences we have with these warning stimuli, or the lack thereof, influences future responding to all similar warning stimuli. In the example above, the behavior used to escape the first warning stimuli (sounds and lights of the fire drill) is reinforced only because a history of consequences does not include the presentation of the second, more aversive stimulus (the fire). Even further, the behavior of escaping the aversive sound and flashing lights is reinforced first, and then the manner in which you escaped more quickly (a shorter route out of the building) or effortlessly (through a door instead of a window) is potentially reinforced second, regardless of whether the fire occurs. As a result of this contingency, individuals create rules about the function and importance of these warning stimuli, and the likelihood of the potential second, more aversive stimulus (it is also possible that the functions of the stimuli in a relational frame change as a result of these experiences in a manner consistent with Relational Frame Theory; Barnes-Holmes, Hayes, Dymond, & O'Hora, 2001).

When an individual responds appropriately to warning stimuli and then receives information that a negative consequence was not present (i.e., that a drill took place), the warning stimuli lose their function to signal "danger." Likewise, if a worker responds to these warning stimuli and does not avert a negative consequence, or does not receive reinforcers for responding appropriately, the behavior will be placed on extinction (resulting in reduced probability

[1]We base our analyses on the definition of Blakely and Schlinger (1987) that rules are "function-altering contingency-specifying stimuli" (p. 183), which is further summarized by Agnew and Redmon (1993) who explain that "rules influence behavior by changing the function of other stimuli; and it is those other stimuli, whose function has been altered, that directly control behavior" (p. 68).

of that behavior in the future).[2] That is, one has engaged in safety behavior without a consequence other than the negative consequences involved in the time and effort to respond. After repeated exposure to the warning stimuli without avoiding the more aversive negative consequence (safety incident), or more direct positive reinforcement (managerial praise), the warning stimuli will lose their discriminative function (i.e., control over behavior that is needed to avert the negative consequence). Unfortunately, these fire drills and individuals' responses to these warning stimuli can generalize to other settings involving aversive sounds and lights to signal danger, including work settings.

However, Bogard et al. (2015) put forth an example where new workers initially responded to warning stimuli but as their tenure grew on the job they stopped responding. This pattern cannot be explained using the analysis suggested above. Escaping an aversive stimulus should lead to an increase in the response that resulted in the escape of the aversive warning signal. Whatever behavior the worker engaged in to escape the warning sound will be reinforced. A worker can fail to follow the process safety procedures and still escape the aversive warning sound (e.g., override the warning signal by entering a computer code without inspecting the equipment). Since the worker escapes the aversive warning sound/lights it is *that* consequence that will control the behavior, and not the potential consequence of averting an explosion or death (see Figure 1). Additionally, the worker does not come into contact with the possible, cumulative negative consequences that are occurring from escaping the warning sound with a nonsanctioned response. Therefore, every time the worker is able to escape the warning signal without following the approved standards, those approved standards become neutral stimuli and appropriate safety responses are replaced by inappropriate, self-devised responses. Alternatively, the worker may generate rules about these relationships that may come to control the response. For example, a rule like "turning off the warning sounds gets rid of it, and nothing bad happens." It would be difficult to devise an experiment to determine whether rules or contingency-shaped behavior are controlling the behavior in any given instance, but a laboratory simulation in which verbal behavior is disrupted under critical conditions might give an indication of the function of that particular behavior.

In the preceding section, we have attempted to explain deviations in worker responses to warning stimuli embedded in process safety. We have mainly made reference to negative reinforcement (escape and avoidance learning) and rule governance. We will now present a case for understanding responses to those stimuli by analogy to habituation.

---

[2]It was questioned if this would be extinction if a fire was never present, but we argue that a worker likely received reinforcement for responding appropriately in a training situation or drill either in that work environment, or a similar context.

Habituation to an eliciting stimulus

Respondent behavior

Habituation is defined as a decline in the magnitude of unlearned responses (e.g., salivation, startle response) due to repeated exposure to an unconditioned stimulus (e.g., food or loud noise; Harris, 1943). When a particular stimulus is presented repeatedly, the stimulus loses its capacity to elicit that response, thus resulting in response decrements (Davidson & Benoit, 1998). In the previous example of workers who behave to escape aversive auditory and visual stimuli (warnings), these warning stimuli may lose their capacity to elicit the startle response after repeated exposure. What this means for process safety is that warning signals emitted repeatedly within a given time period due to a malfunction, or a perceived malfunction, lose their ability to elicit the startle response to "danger."

Operant behavior

Long-term habituation is likely a more prevalent phenomenon in process safety than habituation with respondent conditioning. Long-term habituation is a characteristic of habituation that has been observed with operant responding and is described in the basic research literature by its effects on behavior (McSweeney, 2004). It involves decrements that (a) are learned, (b) are gradual with each subsequent exposure to the stimulus, and (c) occur over days or weeks (McSweeney; McSweeney & Swindell, 2002).

Long-term habituation can best be explained by analogy to the behaviors of nonhumans responding to predator stimuli. Predator stimuli are conditioned stimuli for an animal that learns the sight and odor of its predators, and responds to the predator with conditioned behaviors such as visual scanning and exploratory activity. Species survival is a key contributor in the evolutionary process whereby animals that appropriately respond to threatening stimuli have a greater chance of protecting their life and the lives of their offspring. Predator attack is a situation in which defenses are useful and one gains an advantage by responding with anxiety, fight/flight response, or avoidance learning (Nesse, 2005). However, not all stimuli of predators are associated with attacks and animals must learn to discriminate between threatening versus nonthreatening stimuli.

Animals that fail to discriminate (and as a result of the failed discrimination) respond to all stimuli present in their environment. But animals that habituate to nonthreatening stimuli (e.g., a sleeping lion, or a recently fed leopard) conserve more energy, thereby gaining an advantage and increasing their chances of survival (McSweeney & Swindell, 1999). In terms of process safety, a worker learns to discriminate between "threatening" and "nonthreatening" auditory/visual stimuli, and habituates to these stimuli over time due to repeated presentations over days, weeks, and months. Therefore, with each presentation of a warning alarm, exploratory activity to investigate the cause of an alarm decreases to conserve energy since a response to "nonthreatening" stimuli is not appropriate.

Furthermore, research on discrimination between threatening and neutral stimuli has revealed that nonhumans (e.g., monkeys) demonstrate reactivity (i.e., display fear-related responses) to a predator stimulus after initial exposure. After repeated presentation of the stimulus without contact or attacks (i.e., without a consequence), they habituate to that stimulus (i.e., responding decreased rapidly and gradually; Ellard, 1996; Glaudas, Winne, & Fedewa, 2006; Hall, Bradshaw, & Robinson, 2002; Kemble & Bolwahnn, 1997). In comparison, a worker reacts to the presence of a warning stimulus in the environment, but over time could habituate to the signal if the signal was not paired with a consequence to maintain the discriminative function properties that control responses of survival. This would help explain why there are workers who slightly deviate each time they engage or respond, as opposed to those who fully deviate the first time they engage in the behavior (Bogard et al., 2015).

Furthermore, workers are indeed responding to the threat by eliminating the reason for the auditory and visual warning signals. In the case of process safety, the problem becomes complex because the secondary threat (i.e., the industrial accident) is simultaneously present, and that threat does not necessarily require a second response if there is a malfunction in the warning system (see Figure 1, decision for Response #2 and the consequences of both response choices). Instead, the first response may or may not minimize/eliminate the secondary threat. This may help to explain the differences in responses between workers. As noted by Nesse (2005), "expressing such responses will give a net pay off when the cost of the defense is less than the benefit (amount of harm reduction)" (p. 94). "If repeated exposure to a danger or a cue of a danger is followed by no actual harm, or easily avoidable threat, then the response threshold can safely be raised or the intensity of the response can be reduced" (Nesse, p. 101).

"Normalization of deviance" is theoretical framework that also warrants consideration in this context (Vaughan, 1996). In short, normalization of deviance is an organizational theory about why deviations from protocols (e.g., safety protocols) are gradually reinforced within an organization until a culture of deviance exists, although those deviations seem dangerous or unethical to persons outside that organization. Vaughan used the NASA Challenger explosion to illustrate her theory, as evidence after the fact suggested that deviations from safety standards were disregarded company-wide. In fact, new verbal rules about acceptability of deviations appear to have been generated in the course of discovery of those very same deviations.

These deviations were then not identified until a catastrophic event occurred, as there were no detrimental consequences for the deviance up to that point.

In the example in the preceding paragraphs, it is possible that rule statements about the lack of consequences for simply turning off alarms may come to be reinforced organization- or department-wide to effect normalization of deviance. When these rules begin to control actual behavior in critical situations, alarms are turned off and the likelihood of a catastrophic event is increased.

Risk discounting

Risk discounting is another approach to studying behavior in risky situations using different parameters of safety-related environmental variables to predict risky choices.[3] As discussed above, a multitude of possible stimuli may be affecting the actual response engaged in by the worker. For example, consequence-based variables such as the likelihood of incident or the severity of the consequence may predict risky behavior. Task-related variables, such as the amount of effort required to engage in safe behavior (i.e., response cost; Sigurdsson, Taylor, & Wirth; 2013) and production pressures may also affect choices.

Sigurdsson et al. (2013) exposed undergraduate students to a computer-based simulation construction task with a series of choice scenarios. For each safety choice, participants were presented with two hypothetical scenarios that differed in a parametric fashion along two dimensions: the working height for the construction task and the effort associated with retrieving and donning a safety harness. Participants were then instructed to choose the scenario in which they were more likely to wear the safety harness: (a) a scenario in which participants were working at less height and with less effort and (b) another scenario in which participants were working at more height and more effort was associated with donning the safety harness. Each participant was exposed to over 400 choices, and choice patterns revealed that participants were more likely to make risky choices at greater heights as the effort to retrieve and don a safety harness increased.

Discounting of the value of consequences has been studied extensively in behavior analysis (Green & Myerson, 2004), and results suggest that both non-verbal animals and humans can be sensitive to consequence parameters such as delay and probability of occurrence. For example, pigeons and rats are generally more inclined to choose a smaller immediate edible reinforcer than a larger delayed one. Under certain circumstances, humans are also more likely to choose immediately available smaller rewards then larger delayed ones. This finding may be explained by the fact that delivery of a delayed consequence is not guaranteed,

[3]According to Oxford dictionary, a Choice (1889) is "The act of choosing; preferential determination between things proposed," and a Decision (2015) is "The action, fact, or process of arriving at a conclusion regarding a matter under consideration."

so it may be more advantageous from an evolutionary perspective to eat food that is immediately available rather than risking to miss entirely out on food (i.e., you are less likely to find the food again if you wait).

In a similar vein, humans are more likely to choose a small reward that requires less effort than a larger reward that requires a larger effort to obtain (Botvinick, Huffstetler, & McGuire, 2009). The findings of Sigurdsson et al. may perhaps be explained in a similar fashion. That is, the choice of rather wearing a harness under low-risk situations than high-risk situations becomes more likely as the effort increases in the high-risk situations because the consequence of falling down from heights and getting injured is in general fairly unlikely.

In terms of process safety, this finding may have implications for the response cost associated with different safety choices at critical time points. For example, a warning stimulus (e.g., chemical leak alarm) should set the occasion for follow-up in most cases where an operator would investigate equipment to determine why the warning was set off. If the choice of investigating involves considerable effort (e.g., leaving the control tower, conducting visual inspections of machinery, conducting on-site tests, etc.) it is possible, in light of the findings of Sigurdsson et al., that the operator will choose the alternative of doing nothing and turning off the alarm.

Risk discounting appears to be a potential avenue of research for behavior analysts interested in process safety. A number of further variables seem ripe for study, including probability of negative outcome, time pressures, and various safety culture variables. It is also possible that a history of false alarms may influence choice responding (see discussion on habituation above). Future research could, for example, focus on experimentally manipulating such a history and later conduct tests of risk discounting to observe if a larger number of experiences with false alarms subsequently results in riskier choices. Moving from hypothetical situations to actual work settings would also be a natural step forward in this line of research.

Consequence dimensions

Prospect theory

Another perspective for understanding deviations in process safety involves various other dimensions of consequences of worker behavior. In recent years, empirical research on decisionmaking under risk has produced a body of knowledge that should be of value to those who seek to understand and improve risk management decisions in process safety.[4] The prospect theory of decision making

[4]The language of prospect theory is based in cognitive science constructs. In the following section, we use terms borrowed from prospect theory, but are fully aware of their mentalistic implications. We believe this line of research is relevant to behavioral safety, as it involves carefully and parametrically manipulated response and consequence dimensions. The methodology and results can hence and be easily reconciled with behavioral explanations of behavior.

(Kahneman & Tversky, 1979) suggests that loss aversion may be a dimension of consequences that influences deviations in process safety. According to Kahneman and Tversky, individuals are more likely to behave in ways that result in avoiding a loss rather than gaining a valued consequence. Furthermore, individuals are more likely to make risky choices when loss is certain.

If this component of decision making was applied to our example of process safety, it may help to explain why employees are more sensitive to losses in time or energy, versus the unforeseen gains of finding a chemical buildup and averting a potential negative consequence (see Figure 1, Response #2). Interestingly, research has revealed that individuals are more likely to compare losses against losses (not necessarily losses against gains), because those consequence values are on the same scale (McGraw, Larsen, & Kahneman, 2010). This becomes important when looking at process safety because researchers have found that "when gains and losses are judged separately, but by the same person and on the same scale ... loss aversion is evident but weaker." (McGraw et al., p. 1443). This may mean that some individuals may be comparing losses of time, energy, and production against the losses that occur from a chemical buildup or an explosion, and the potential gains of positive consequences never factor into their weighted decision making. Individuals that are only comparing losses to losses are likely factoring the consequence dimensions (in Figure 1 of the contingency diagram), the losses that have a higher severity (explosion, production stoppage that results if you return to work and do not inspect equipment) appear to have consequence dimensions that are unknown, uncertain, and unlikely. This is contrasted to the other losses (time, energy, production reductions that result from inspecting the equipment) that appear to have consequence dimensions that are known, certain, and likely, and are hence theoretically more likely to influence behavior. Therefore, this may explain why workers are more averse to a loss of time, energy, and production instead of a more aversive loss of explosion and death.

When specifically looking at the consequence dimensions of decision making, prospect theory research has found that a high probability of gains leads to being risk-averse, whereas a high probability of losing leads to risk seeking; conversely, a low probability of gaining leads to risk seeking and a low probability of losses leads to being risk-averse (Kahneman & Tversky, 1979). For example, individuals will make a riskier decision when the decision results in a consequence of losing more (−4 points) with an 80% probability because there remains a 20% probability of losing zero, versus making the "safe" decision of losing a smaller amount (−3 points) with 100% probability.[5] This decision making is categorized as risk seeking. However, individuals will make a "safe" decision when there is a 100% probability they will gain a smaller amount (3 points) as opposed to making the riskier

[5]Participants in this study were aware of the probabilities and consequences of their decision making prior to making the decision and they did not encounter feedback to shape their decisions since it was a one-opportunity decision; this is common in description-based decision research.

decision that results in a consequence of gaining more (4 points) with an 80% probability because there remains a 20% probability of receiving zero points; this decision making is termed risk aversion (Kahneman & Tversky; in Barron & Erev, 2003).

When applying this theory of decision making to process safety we see that there are not many gains (reinforcers, rewards) that are received for responding safely. That is, there is a low probability workers will obtain a reward for safe behavior (see Figure 1, Responses 2 and 3), but a high probability they will lose something of value (time, energy, production during Responses 2 and 3) and both of these scenarios lead to risk seeking decision making.

When comparing the distribution and probability of gains and losses for engaging in safety behavior versus at-risk behavior, behavior analysts have covered the topic of probability; the most familiar source is that offered by Daniels and Bailey (2015) who suggest various dimensions of consequences that factor in maintenance of behavior. Consequences that are certain and immediate outweigh the less probable and sometimes delayed losses associated with at-risk behavior. For example, with the Deepwater Horizon accident, it is possible that one contributing factor was that turning off the warning signals provided an immediate, certain, positive consequence in the form of quiet surroundings and sleep, whereas the negative consequence of a disaster by not responding to the alarm was delayed and uncertain. These decision-making probabilities are then inverted when using safe behavior associated with uncertain and delayed consequences. Time and cost restraints, and associated managerial pressure, also played a role in the Deepwater Horizon accident, so that acting properly on safety signals can be said to have been punished, and deviating from safety protocols to have been reinforced. Future studies in process safety could focus on management behavior in process safety, and look for ways to improve their use of safety-related feedback and praise. Cooper (2006), for example, incorporated measures of management support in a behavioral safety application in a paper mill, and found that as management's support behaviors increased, the safety performance of workers increased as well.

Another decision-making theory examines the influence of reinforcement and punishment on decision making when choosing between responses. The reinforcement-based learning model was influenced by Thorndike's law of effect and "assumes that the probability that a certain strategy will be adopted increases when this strategy is positively reinforced" (Erev, Gopher, Itkin, & Greenshpan, 1995, p. 370). Conversely, use of a strategy will decrease when the outcome that follows is aversive (Barkan, Zohar, & Erev, 1998). Furthermore, individuals will frequently choose an action when there is a high probability that action is the best choice (Erev, 1998). Simulations on choice across varying consequences revealed that when the probability of being penalized decreased, risky decisions increased (Barkan, et al.). Data also appeared to indicate that riskier behavior occurs when there is a low probability of a negative outcome, and in the absence of reinforcement, but large negative consequences (e.g., losing points) resulted in fewer risky decisions. Finally, Barkan et al. found that when uncertainty is present in the decision-making process, individuals underestimate the likelihood of a rare event (e.g., catastrophic accident).

This line of research was further developed by Barron and Erev (2003) who examined "small feedback-based decisions" that are characterized, in part, by a problem that is repeatedly presented in resemblant conditions, each choice has similar, but small probabilities and the consequences of each alternative are based on "immediate feedback obtained in similar situations in the past" (p. 216). Essentially, their study on choice was devised to examine when individuals used only the feedback (i.e., the consequence or "payoff" received) they obtained from making a decision (i.e., choosing between two buttons on a computer to simulate "payoffs" of gains and losses based on their selection). Participants were provided the goal to earn as many points as possible, which was connected to a financial payoff at the end of the study (which can be comparable to production demands in process safety). Results from the experiment showed that when decisions produce feedback in the form of consequences, that feedback influences subsequent decisions in similar choice situations, and such feedback resulted in participants underweighting low probability events. Furthermore, individuals were influenced more by the probability of a rare event than by the magnitude of the rare event due to the feedback immediately received after a decision (Barron & Erev, 2003).

It should be noted that a main difference in Barron and Erev's (2003) research from Kahneman and Tversky's (1979) prospect theory pertains to Barron and Erev addressing the role of immediate feedback (i.e., encountering the consequences of the decision) in shaping future decisions in similar situations, and individuals "learning" to evaluate and determine risk (size, type, probability) associated with each choice via feedback from their choices, thus influencing future decision making (i.e., is the risk worth taking). The feedback-based decision-making model of Barron and Erev reflects repeated decisions and outcomes of those decisions, which found that individuals underweight rare events. This experimental model appears to simulate the decision making of an experienced worker.

Conversely, prospect theory derives from a one-time opportunity decision (i.e., the first exposure to a situation) wherein individuals overweight the small probability of a rare event; this appears to simulate the decision making of a novice worker. This suggests that individuals who are more experienced with a situation due to increased number of opportunities with that situation and with encountering the consequences that follow their decision making leads to riskier decision making because consideration is not given to the rare event. Conversely, novice individuals who have limited experience and have not received consequences because a situation is new, are found to give more consideration to rate events (Barkan, et al.; Kahneman & Tversky).

This may help explain why new employees (i.e., novices) give more consideration to the rare event and follow protocols and as experience develops they begin to make riskier decisions in situations in which they previously engaged in "safe" decisions. More specifically, novice employees have no prior experience or information about the consequences and probabilities of their various decisions with a specific situation and thus, their behavior is shaped by immediate feedback and from the consequences of their decisions. Employees hence appear to develop response tendencies with experience to match the "payoff" of certain decisions (Barron & Erev). However, irrespective of experience, loss aversion is a frequent property of decision making and has been repeatedly replicated in the literature (Barron & Erev; Kahneman & Tversky). Researchers could examine this by creating novel stimuli with which individuals have no previous learning history (e.g., those used to study stimulus equivalence), present choice that simulates risk via point gain and loss, and examine individuals' self-report estimates of choice-probability and their subsequent choices after encountering the consequences of their choice. Throughout the experiment as "experience" is developing, the task directions could be tracked to determine how long participants follow communicated directions and when the deviation begins.

Heuristics

The process safety community would also benefit from research that helps mitigate the effects of time pressure, and individuals' possible use of heuristics to find a "good" solution without all the relevant information needed to evaluate the consequences of their decision making. A heuristic is described as "a strategy that ignores part of the information, with the goal of making decisions more quickly, frugally, and/or accurately than more complex methods" (Gigerenzer & Gaissmaier, 2011, p. 454). Gigerenzer and Gaissmaier explain further with the fluency heuristic: "if both alternatives are recognized but one is recognized faster, then infer that this alternative has the higher value with respect to the criterion" (p. 462). It is commonly assumed that heuristics result in less accurate/lower quality decision making, however, according to Gigerenzer and Gaissmaier that is not necessarily true for all contexts. Researchers have found that decision making (with complex decisions) between alternative choices that are difficult to distinguish leads to difficulty in determining the correct/most effective alternative (Lau & Redlawsk, 2001). In these situations, experience did improve the probability of selecting the correct alternative; the opposite was found with inexperienced individuals, likely due to them not understanding the context in which they were making the decisions or being less knowledgeable about the choices (Lau & Redlawsk). If employees are employing heuristics before responding to a warning, in the absence of complete information, evaluation, or guidance, and a negative consequence does not follow (no accident), and reinforcement is provided (escaping aversive sounds and lights, escaping punishment of a potential accident), it would appear that a good strategy for decision making was used in that situation, thus reinforcing the employee's use of a heuristic (from the employee's perspective).

Research has also examined the effects of time on using heuristics and found it to be an influencing variable wherein efficiency becomes more important than accuracy (Payne, Bettman, & Johnson, 1988). That is, individuals "minimize effort by using a diverse set of heuristics, changing rules as contexts and time pressures change" (p. 541). Workers may perceive time pressures based on aversive auditory/visual sounds, pressure from others in the process to set up or fix machinery, and time pressure associated with company production demands. Payne, Bettman, and Johnson found according to their simulations, "it seems important under high time pressure to use a choice strategy that processes at least some of the information about all alternatives as soon as possible" (p. 541). However, while the use of heuristics may be reinforced because the probability of a severe accident is low, it is debatable if heuristic choice strategies should be involved in responding to process safety issues.

Knowing all the relevant information for each alternative would appear critical in process safety and studying factors that increase the probability of avoiding short-cut heuristics seems an area ripe for behavior analysts. Research could examine what types of safety information (e.g., which consequence dimensions) are most important to process first, and if processing only some of that information (or a specific combination of information components) about all the choices leads to "safer" decisions, instead of processing all the information of the first choice then all the information of the second choice. Further, one could examine the use of heuristics with safety while simulating time pressure by having individuals complete production goals simultaneously while making choices. Finally, employees should be trained on the consequences of this decision strategy and how it can be faulty with unintended consequences. For example, Endsley (1995) developed a model known as situation awareness to explain the influence of perceiving and comprehending information when individuals are engaged in decision making, and how faulty decision making can occur under complex environmental factors. Killingsworth, Miller, and Alavosius (2016) provide an extensive behavioral interpretation of situational awareness that accounts for an employee's verbal behavior while assessing and responding to changing contingencies in a complex work environment. In this behavioral interpretation of situational awareness, there is a sequence that an employee goes through to reach a decision, which includes (a) perceiving: does one accurately perceive the environmental stimuli; (b) comprehending: does one understand the meaning and significance of the stimuli; and (c) projecting: predicting potential future events, which leads to a decision that is either correct or not. There are accepted steps or rules for a response to a situation, but if an employee is not able to accurately recall the steps or rules, or their behavior does not come under stimulus control, then an appropriate response (i.e., safe behavior) is compromised (Killingsworth et al., 2016). Described further behaviorally, when presented a stimulus, does the stimulus occasion the correct response (i.e., stimulus control), and in the presence of various stimuli, can the employee discriminate (i.e., conditional discrimination) between these stimuli and respond according to a reinforcement history? Killingsworth et al. (2016) explains this sequence below:

> *'How an individual reacts in a given situation focuses on the observable stimulus-response interactions ... verbal behavior comes under the control of stimuli ... after one emits discriminated responses in the presence of relevant task stimuli, a subsequent verbal statement is emitted and this sets the occasion for concurrent or subsequent responding' ... and responding to stimuli is due to a reinforcement history with those stimuli or that 'the ability to predict is enabled by a history of similar response and stimulus functions in the past' (pp. 308–310).*

Situation awareness is a model referenced frequently in corporations, which explains how workers become unsafe via "complacency," and thus, there is great need for behavior analysts to become familiar with a behavioral understanding of "complacency" and situation awareness to provide better direction to corporations to intervene on another potential source of behavioral variability in process safety.

Behavioral decision theory

Behavioral decision theory attempts to study *how* decisions are made by "tracing" decisions as a person makes them, via information boards (Redlawsk & Lau, 2013). Information boards consist of columns and rows that contain various decision options and the consequence dimensions of each option. The various combinations of the decisions are not presented to the individual at once, but presented as feedback after a selection is made, thus serving as an analysis of decision-making strategies (Redlawsk & Lau; see Mintz, Geva, Redd, & Carnes, 1997 for a diagram of an information board and the dimensions and consequences of each alternative choice). The components that are examined include the choice selected, the duration of "thinking" about an alternative in between selections, and the arrangement of sequential selections (Redlawsk & Lau).

Dynamic information boards are an extension of information boards, which attempt to replicate complex decision making that occur outside a laboratory with more uncertainty, uncontrollable outcomes, and information that develops over time. Such research has found that when the decision-making environment is complex (i.e., processing high volume and a variety of information), individuals simplify their decision-making strategies by avoiding an analysis of consequence dimensions to prevent a conflict between a new, unpleasant piece of information against their preferred decision (Redlawsk, 2004; Redlawsk & Lau). More specifically, learning contradictory information about a preferred decision becomes more "cognitively taxing" (i.e., more information to manage, more stress, and an increase in time and effort) because the positive and negative consequence dimensions have to be considered, making the decision more difficult (Redlawsk). Therefore, it is less "cognitively taxing" to ignore the information that contradicts your preferred decision, or to not seek the contradictory information at all (Redlawsk). Applying this to process safety, it is possible that "cognitively taxing" decision making equates to increased response cost for workers, which may serve as punishment and thus, workers engage in decision making that escapes the aversiveness (i.e., engage in decision making that is easier, quicker, and/or less effortful).

In summary, by employing heuristics and noncompensatory rules during decision making, individuals minimize the amount of information needed to make a decision, either by evaluating only some of the consequence dimensions of each alternative, evaluating only some of the alternatives fully, or ignoring and/or not seeking information about an alternative choice. Behavioral decision-making research has shown that all the consequences and consequence dimensions of a decision are not weighed and evaluated prior to a decision by an individual (Lau, 1995; Redlawsk & Lau).

Employing the use of dynamic information boards in laboratory research to examine safety decisions, and evaluating the consequences and time spent on contradictory safety choices (e.g., choices that are known to lead to more loss), may provide an understanding in an area of behavior analysis that is rarely examined. For example, participants could be provided a safety scenario, and asked to make a choice and explain the rationale for their choice, but the individual would then be presented contradictory information that their choice is "risky" (regardless of their choice, contradictory information would always be provided to participants). Analysis could focus on examining how long an individual will continue to investigate contradictory information about their choice versus ignoring the contradictory information and continuing to select "risky" choices (see *confirmation bias* and the *backfire effect* literature for a more in-depth review of these topics); and the consequences of their choices could mimic the naturally occurring consequences typically received for unsafe work behavior. This research could further help our understanding of verbal behavior during decision making and situation awareness.

Discussion

Given this review of the selected perspectives on behavioral variability in process safety, researchers and practitioners now have an opportunity to study some variables that affect employees' behavior as it relates to process safety. To minimize extinction of appropriate process safety responses to aversive warning stimuli, practitioners could implement praise and reinforcement for sanctioned process safety responses to off-set natural reinforcement received from the nonsanctioned responses. Furthermore, given that reinforcement is delivered for the first response (escaping the aversive auditory/visual stimuli), another contrived reinforcer needs to be introduced into the environment for the second and third responses of avoiding the unsafe condition and using the appropriate, sanctioned response. Additionally, if performance decrements are due to long-term habituation, then a change to the stimulus should disrupt the extinguished response (Aoyama & McSweeney, 2001). Neese (2005) notes that systematic research on varying intensities of signals is scarce, and therefore, this is a potential avenue for behavioral safety researchers. The safety community would benefit from research that examines the effects of randomly rotating the sounds produced by the machines' technology to lessen habituation to the warning signals, either by introducing a sliding scale of warning signals or delivering the signals in no discernable fashion to minimize workers creating rules. For example, one could examine the effects of adhering to process safety standards by changing the pitch and duration of the warning signals much like the differences between fire trucks and ambulances from different regions in the world.

Researchers may find value in investigating the rules that are generated around human behavior in process safety, such as those that are illustrative of normalization of deviance (Vaughan, 1996) and management practices. Daniels and Bailey (2015) advocate for the addition of immediate managerial consequences to support important organizational behavior, and expanding Cooper's (2006) work to process safety settings seems well suited.

Finally, researchers can expand decision-making research to process safety to simulate decision making that occurs in workplace settings to determine if the probability of being penalized influences risky decision making, the influence of small and large gains and losses on process safety behavior, and expanding the work of Sigurdsson et al. (2013) on increased effort with respect to critical process safety behaviors. Finally, there is great potential in investigating the behavioral interpretation of situational awareness (Killingsworth et al., 2016) to examine faulty decision making in dynamic environments where conditional discrimination is frequent. Decision theory holds potential for our field, since it examines decision strategy, behavior, consequences of behavior, and value of consequences.

References

Agnew, J. L., & Redmon, W. K. (1993). Contingency specify stimuli: The role of "rules" in organizational behavior management. *Journal of Organizational Behavior Management, 12* (2), 67–76. doi:10.1300/J075v12n02_04

Aoyama, K., & McSweeney, F. K. (2001). Habituation contributes to within- session changes in free wheel running. *Journal of the Experimental Analysis of Behavior, 76*, 289–302. doi:10.1901/jeab.2001.76-289

Barkan, R., Zohar, D., & Erev, I. (1998). Accidents and decision making under uncertainty: A comparison of four models. *Organizational Behavior and Human Decision Processes, 74*(2), 118–144. doi:10.1006/obhd.1998.2772

Barnes-Holmes, D., Hayes, S., Dymond, S., & O'Hora, D. (2001). Multiple stimulus relations and the transformation of stimulus functions. In S. C. Hayes, D. Barnes-Holmes, & B. Roche (Eds.), *Relational frame theory: A post-Skinnerian account of human language and cognition* (pp. 51–72). New York, NY: Plenum Press.

Barron, G., & Erev, I. (2003). Small feedback-based decisions and their limited correspondence to description-based decisions. *Journal of Behavioral Decision Making, 16*, 215–233. doi:10.1002/(ISSN)1099-0771

Blakely, E., & Schlinger, H. (1987). Rules: Function-altering contingency-specifying stimuli. *The Behavior Analyst, 10*(2), 183–187. doi:10.1007/BF03392428

Bogard, K., Ludwig, T. D., Staats, C., & Kretschmer, D. (2015). An industry's call to understand the contingencies involved in process safety: Normalization of deviance. *Journal of Organizational Behavior Management, 35*(1–2), 70–80. doi:10.1080/01608061.2015.1031429

Botvinick, M. M., Huffstetler, S., & McGuire, J. T. (2009). Effort discounting in human nucleus accumbens. *Cognitive, Affective, & Behavioral Neuroscience, 9*, 16–27. doi:10.3758/CABN.9.1.16

Choice. (1889). *Oxford English dictionary online*. New York, NY: Oxford University Press.

Cooper, D. M. (2006). Exploratory analyses of the effects of managerial support and feedback consequences on behavioral safety maintenance. *Journal of Organizational Behavior Management, 26*(3), 1–41. doi:10.1300/J075v26n03_01

Daniels, A. C., & Bailey, J. S. (2015). *Performance management: Changing behavior that drives organizational effectiveness*. Atlanta, GA: Performance Management Publications.

Davidson, T. L., & Benoit, S. C. (1998). Learning and eating. In W. O'Donohue (Ed.), *Learning and behavior therapy* (pp. 498–517). Needham Heights, MA: Allyn & Bacon.

Decision. (2015). *Oxford English dictionary online*. New York, NY: Oxford University Press.

Deepwater Horizon Study Group. (2011). *Investigation of the Macondo well blowout disaster*. Retrieved from http://ccrm.berkeley.edu/pdfs_papers/bea_pdfs/dhsgfinalreport-march2011-tag.pdf

Ellard, C. G. (1996). Laboratory studies of antipredator behavior in the Mongolian gerbil (*Meriones unguiculatus*): Factors affecting response attenuation with repeated presentations. *Journal of Comparative Psychology, 110*, 155–163. doi:10.1037/0735-7036.110.2.155

Endsley, M. R. (1995). Toward a theory of situation awareness in dynamic systems. *Human Factors, 37*(1), 32–64. doi:10.1518/001872095779049543

Erev, I. (1998). Signal detection by human observers: A cutoff reinforcement learning model of categorization decisions under uncertainty. *Psychological Review, 105*, 280–298. doi:10.1037/0033-295X.105.2.280

Erev, I., Gopher, D., Itkin, R., & Greenshpan, Y. (1995). Toward a generalization of a signal detection theory to N person games: The example of two person safety problem. *Journal of Mathematical Psychology, 39*, 360–375. doi:10.1006/jmps.1995.1034

Gigerenzer, G., & Gaissmaier, W. (2011). Heuristic decision making. *Annual Review of Psychology, 62*, 451–482. doi:10.1146/annurev-psych-120709-145346

Glaudas, X., Winne, C. T., & Fedewa, L. A. (2006). Ontogeny of anti-predator behavioral habituation in cottonmouths (*Agkistrodon piscivorus*). *Ethology, 112*, 608–615. doi:10.1111/eth.2006.112.issue-6

Green, L., & Myerson, J. (2004). A discounting framework for choice with delayed and probabilistic rewards. *Psychological Bulletin, 130*, 169–792. doi:10.1037/0033-2909.130.5.769

Hall, S. L., Bradshaw, J. W. S., & Robinson, I. H. (2002). Object play in adult domestic cats: The roles of habituation and disinhibition. *Applied Animal Behavior Science, 79*, 263–271. doi:10.1016/S0168-1591(02)00153-3

Harris, D. (1943). Habituatory response decrement in the intact organism. *Psychological Bulletin, 40*, 385–422. doi:10.1037/h0053918

Kahneman, D., & Tversky, A. (1979). Prospect theory: An analysis of decision under risk. *Econometrica, 47*, 263–292. doi:10.2307/1914185

Kemble, E. D., & Bolwahnn, B. L. (1997). Immediate and long-term effects of novel odors on risk assessment in mice. *Physiology & Behavior, 61*, 543–549. doi:10.1016/S0031-9384(96)00499-4

Killingsworth, K., Miller, S. A., & Alavosius, M. P. (2016). A behavioral interpretation of situation awareness: Prospects for Organizational Behavior Management. *Journal Of Organizational Behavior Management, 36*, 301–321. doi:10.1080/01608061.2016.1236056

Lau, R. R. (1995). Information search during an election campaign: Introducing a process-tracing methodology for political scientists. In M. Lodge, & K. M. McGraw (Eds.), *Political judgment: Structure and process*. Ann Arbor, MI: University of Michigan Press.

Lau, R. R., & Redlawsk, D. P. (2001). Advantages and disadvantages of cognitive heuristics in political decision making. *American Journal of Political Science, 45*, 951–971. doi:10.2307/2669334

McGraw, A. P., Larsen, J. T., & Kahneman, D. (2010). Comparing gains and losses. *Psychological Science, 21*(10), 1438–1445. doi:10.1177/0956797610381504

McSweeney, F. K. (2004). Dynamic changes in reinforcer effectiveness: Satiation and habituation have different implications for theory and practice. *The Behavior Analyst, 27*(2), 171–188.

McSweeney, F. K., & Swindell, S. (1999). Behavioral economics and within-session changes in responding. *Journal of the Experimental Analysis of Behavior, 72*, 355–371. doi:10.1901/jeab.1999.72-355

McSweeney, F. K., & Swindell, S. (2002). Common processes may contribute to extinction and habituation. *The Journal of General Psychology, 129*, 364–400. doi:10.1080/00221300209602103

Mintz, A., Geva, N., Redd, S., & Carnes, A. (1997). The effect of dynamic and static choice sets on political decision making: An analysis using the decision board platform. *The American Political Science Review, 91*(3), 553–566. doi:10.2307/2952074

Nesse, R. M. (2005). Natural selection and regulation of defenses: A signal detection analysis of the smoke detector principle. *Evolution and Human Behavior, 26*, 88–105. doi:10.1016/j.evolhumbehav.2004.08.002

Occupational Safety and Health Administration. (2000). *Process safety management; OSHA 3132 (reprinted)*. Retrieved from https://www.osha.gov/Publications/osha3132.html

Payne, J. W., Bettman, J. R., & Johnson, E. J. (1988). Adaptive strategy selection in decision making. *Journal of Experimental Psychology: Learning, Memory, and Cognition, 14*(3), 534–552.

Redlawsk, D. P. (2004). What voters do: Information search during election campaigns. *Political Psychology, 25*(4), 595–610. doi:10.1111/pops.2004.25.issue-4

Redlawsk, D. P., & Lau, R. R. (2013). Behavioral decision-making. In L. Huddy, D. O. Sears, & J. S. Levy (Eds.), *The oxford handbook of political psychology* (2nd ed.). New York, NY: Oxford University Press.

Sigurdsson, S. O., Taylor, M. A., & Wirth, O. (2013). Discounting the value of safety: Effects of perceived risk and effort. *Journal of Safety Research, 46*, 127–134. doi:10.1016/j.jsr.2013.04.006

Vaughan, D. (1996). *The Challenger launch decision: Risky technology, culture, and deviance at NASA*. Chicago, IL: University of Chicago Press.

A Behavioral Interpretation of Situation Awareness: Prospects for Organizational Behavior Management

Kenneth Killingsworth, Scott A. Miller, and Mark P. Alavosius

ABSTRACT
Situation awareness (SA) is a construct used in human factors research and application. It is typically employed in the design of equipment to facilitate rapid and adaptive responding in dynamic and high-risk environments. Although the theory backing the SA concept is not entirely compatible with behavioral philosophy, components of the analysis and measures employed in SA work can benefit researchers and practitioners in Organizational Behavior Management (OBM). The present discussion includes (a) the definition and context for SA, (b) a behavioral interpretation of SA, (c) the assessment tools used in SA work, and (d) the relevance of SA to behavioral research. This discussion is pertinent to behavior analysts who work in industries where SA terminology is used and seek analytic tools to guide the design of effective interventions.

In demanding and dynamic work situations, performers are required to respond to constantly changing variables to maintain safety and produce quality work products. Managing process controls in oil refining, driving, flying an airplane, operating heavy equipment, and managing manufacturing operations are examples of jobs that require quick and decisive action or risk failure with potentially catastrophic results. Behavior analysts working in these contexts bring powerful tools to enhance the prediction and control of critical behavior classes and outcomes. Human factors professionals, who often work in the same contexts, but use different analytical tools, may produce outcomes similar to those produced by behavior analysts. One of these tools is the construct situation awareness (SA; Endsley, 1995a), which refers to a performer's ability to assess and respond in dynamic contexts. Currently, SA is receiving considerable attention in industries where automation replaces manual work tasks, and the performer monitors rather than operates equipment (e.g., Behzad & Mann, 2014; Roberts, Flin, & Cleland, 2016; Stevens-Adams et al., 2015; van de Merwe, Oprins, Eriksson, & van der Platt, 2012).

Behavior analysts who are familiar with the SA concept and how it is used to analyze performance in dynamic situations may be able to make use of some of the tactics employed by human factors researchers. Interpreting these tactics within behavioral philosophy may create more parsimonious analyses and encourage the development of new behavioral technologies. The purpose of this paper is to describe SA to behavior analysts, highlighting (a) the definition and context for SA, (b) a behavioral interpretation of SA, (c) SA assessment tools, and (d) behavioral research and SA.

Definition and context for SA

Multiple definitions for SA have been suggested throughout the development of the SA concept. Table 1 presents SA definitions based on conceptual papers and research. We focus on Endsley's definition because it is more developed than other models, captures common features across different approaches to SA, and is considered a representative publication in the human factors community (Wickens, 2008). Endsley (1995a) defines SA as a state of knowledge that results from progressing through three phases: the perception of the task-relevant stimuli, comprehension of their meaning in relation to the task, and the projection of the status of these stimuli in the future.

The three levels of SA

The core components of Endsley's SA model are perception, comprehension, and projection. Perception (Level 1) entails contacting stimuli relevant to the particular task. For instance, a driver may see the "fuel low" or "check engine" light while driving, or a drilling equipment operator may glance at the oil pressure or temperature gauge while operating. The focus in Level 1 is the quality of the information being perceived (i.e., stimuli) because inaccurate or missing information will ultimately affect the other levels of SA and subsequent performance.

Comprehension (Level 2) entails understanding the meaning of the elements that were perceived at Level 1. According to Endsley (1995a), the performer uses the perceived elements from Level 1 to form patterns or mental models of the current situation. Endsley uses the example of a pilot comprehending the "meaning" of three enemy aircraft in a specific formation (i.e., it is an attack formation). The ability to comprehend meaning from environmental stimuli is purportedly one of the factors that separates a novice from an expert pilot (Endsley, 1995a).

The third level of SA is predicting future situations, or projecting the status of relevant task stimuli in a future situation. A performer will attempt to predict a future state of events based on his comprehension of the situation. Good information (Level 1) leads to good comprehension (Level 2), and good comprehension leads to an increased capability to predict possible hazards

Table 1. Definitions of Situation Awareness (SA) and Relevant Behavioral Terms

Definition	Relevant behavioral terms	Source
SA is the perception of the elements in the environment within a volume of time and space, the comprehension of their meaning, and the projection of their status in the near future.	Stimulus control, tact, relational responding, conditional discrimination, entailment	Endsley (1995a)
SA is an abstraction that exists in our minds. The depiction or representation of the elements of a situation can be manipulated by the designer. The understanding of those information elements can be manipulated by proper training. The awareness and interpretation of the meaning of that information by the operator can be improved by practice.	Fluency, stimulus equivalence, intraverbal, stimulus-stimulus pairing, tact, relational network, transformation of stimulus functions	Billings (1995)
A pilot's (or aircrew's) continuous perception of self and aircraft in relation to the dynamic environment of flight, threats, and mission, and the ability to forecast, then execute tasks based on that perception ... it encompasses the individual's experience and capabilities, which affect the ability to forecast, decide, and then execute. SA represents the cumulative effects of everything an individual is and does as applied to mission accomplishment.	Relation responding, tact, historical analysis, generativity, generalization, deictic relation, mutual/combinatorial entailment, recombinative generalization	Carol (1992)
One's ability to remain aware of everything that is happening at the same time and to integrate that sense of awareness into what one is doing at the moment.	Deictic, relational frame, rule-governance,	Haines and Flateau (1992).
SA is adaptive, externally-directed consciousness that has as its products knowledge about a dynamic task environment and directed action within that environment ... SA, as we define it, is a specific brand of adaptation. Adaptation is the process by which an agent channels its knowledge and behavior to attain goals, tempered by the conditions and constraints imposed by the task environment.	Rule-governance, tact, discriminative stimuli, learning channels, stimulus/response classes, interdependent responses, non-linear analysis, response blocking, conditional discrimination, fluency	Smith and Hancock (1995)
SA is the continuous extraction of information about a dynamic system or environment, the integration of this information with previously acquired knowledge to form a coherent mental picture, and the use of that picture in directing further perception of, anticipation of, and attention to future events.	Tact, stimulus control, joint control, conditional discrimination, historical analysis, relational responding	Wickens (1995)

(Level 3). For example, projection is important in air traffic control because the current status of aircraft on the ground will affect the flight patterns of incoming planes. If an air traffic controller accurately projects that a crash is likely, commands can be given to alter the patterning of incoming flights to avoid a catastrophe (e.g., see Durso et al., 1998).

Individually, the three levels above are not considered SA. SA is the state of knowledge that is gained by transitioning through the three levels. In other words, it is the outcome of a process that begins with exposure to information, which is mentally processed and comprehended, and this comprehension allows the performer to predict a future state of events. Subsequent decisions and performance are not included as part of SA, though they are affected by SA, mental faculties, and contextual factors.

The mental mechanisms of SA

The SA model relies on hypothetical mental mechanisms within its explanatory framework. For example, Endsley (1995a) stated,

> ... individuals vary in their ability to acquire SA, given the same data input. This is hypothesized to be a function of an individual's information-processing mechanisms ... the individual may possess certain preconceptions and objectives that can act to filter and interpret the environment in forming SA. (p. 35)

The mental mechanisms in SA affect every aspect of an event from perception to action. These mechanisms relate external stimuli to information stored in the brain in order to create mental models of a situation. For example, task goals and objectives are stored in long-term memory, and this is accessed when a performer perceives (Level 1) the relevant external stimuli. In any given situation, the mental mechanisms perform the bulk of the work by accepting, processing, relating, and storing information for immediate and later use.

Additionally, Endsley uses the term "non-human contextual factors" to separate external stimuli from a performer's mental model of a situation. That is, the non-human contextual factors serve as inputs for the mental mechanisms to create a mental model. For example, backup cameras and rearview displays on vehicles allow drivers to see objects better behind the vehicle. When a driver puts the car in reverse, the image on the LCD screen and the auditory beep are non-human contextual stimuli that are processed by internal mechanisms that output a mental model of the self, vehicle, and the surrounding environment. Endsley further elaborates on how the mental mechanisms process external stimuli (e.g., the functions of attention, schema, schemata, short-term and long-term memory storage).

To be sure, the hypothetical and mentalistic constructs espoused by Endsley are not consistent with behavioral philosophy. Traditionally, behavior analysts have avoided these constructs because their existence is purely hypothetical, and attempting to specify their functions does not add explanatory or predictive power to the analysis of behavior. Discourse on these constructs also detracts from a description of the relationships between the observable things and events involved in a performance (Skinner, 1953). Thus, the mentalistic constructs described in SA will not be critically analyzed in the present paper. Rather, we focus on the context in which SA has been employed and the behavioral events within these contexts.

The context for SA

SA is commonly used in organizational training, safety, interface design, and human factors analyses. Since its beginnings, the term has been used in an analysis of the putative sequences by which pilots and soldiers quickly and accurately respond in challenging and dynamic situations (e.g., an engine failure, a surprise ambush, in situ changes to a plan of execution). Current development of SA has expanded to include performance and safety in complex settings such as health care, roadways, offshore drilling, process manufacturing, and nuclear power. The following scenario is adapted from health care research that uses SA concepts as part of an assessment for nursing skills in emergency situations (i.e., Cooper et al., 2010). This scenario highlights the relationship between levels of SA and real-world events.

Emergency room nurses work in a dynamic and unpredictable environment. A patient may present as stable yet can rapidly deteriorate without warning; this creates a time-sensitive situation. When a patient unexpectedly deteriorates, the charge nurse (CN) is responsible for managing the crisis. The CN must first perceive the relevant stimuli (Level 1). This includes coming into contact with the patient, visual cues from the patient (e.g., skin color, breathing rate), information about the patient provided by staff and anyone accompanying the patient, and any data from medical equipment applied by other staff (e.g., heart rate, blood pressure, oxygen saturation). Also included are other external stimuli, such as the proximity of personal protective equipment, proximity of other staff, and the availability of specialized emergency equipment. Exposure to these and other stimuli allow the CN to extract meaning (Level 2, comprehension) from the situation and begin creating a mental model. In other words, the CN mentally assembles the facts. For example, a patient presenting with a poorly bandaged and seeping wound, patches of discolored skin, rapid heart rate, and disorientation leads the CN to conclude that the patient may be in septic shock. The CN is also "aware" of the medical equipment and staff available to assist in the crisis. After comprehending this Level 1 information, the CN can project the status of the situation in the near future (Level 3, projection). That is, the CN can predict whether the available staff and equipment will likely be enough to successfully stabilize the patient, and the CN's subsequent decisions are based on this projection.

With progress through each level, the CN's state of knowledge of the situation is updated (i.e., SA is updated). Multiple variables change as the CN continues to coordinate staff, administer medical procedures, and monitor the patient. According to the SA framework, to be effective, the CN's mental model must change to accept, comprehend, and project the changing situation. If the information is incomplete, or if either the comprehension or projection is errant, the mental model deviates from the real world. This discrepancy puts task outcomes at risk, and in this case the patient's health is at stake.

The emergency room scenario highlights the use of the SA framework to partition a sequence of events in dynamic situations. This framework continues to serve as an analytical tool to investigate a variety of phenomena: the effects of training (e.g., Soliman & Mathna, 2009), the design of equipment (e.g., Baumgartner, Gottesheim, Mitsch, Retschitzegger, & Schwinger, 2010), post-hoc accident analyses (e.g., Roberts, Flin, & Cleland, 2015), process safety in drilling operations (Roberts et al., 2016), and the effects of automation on performance (e.g., Omnasch, Wickens, Li, & Manzey, 2014). The SA framework has been effectively used to identify interventions that enhance performance, and it continues to expand into industries where behavior analysts work (e.g., health care, oil and gas, transportation). Thus, a behavioral interpretation of the SA phenomena is warranted.

A behavioral interpretation of SA

The pragmatic stance of SA aligns with the focus of behavior analysis, which is prediction and control (Skinner, 1953, 1969, 1974). The decision to investigate SA depends on whether analyzing a sequence of events as suggested by the SA framework, and the behavioral phenomena within those sequences, affords prediction and control over behavior, processes, and outcomes. For behavior analysts, the foray into SA is potentially fruitful because SA researchers have focused on the associations between environmental events and observable outcomes that are meaningful in a variety of industries.

The SA constructs might be seen as convenient labels for the behavioral events involved in a masterful performance. The second column in Table 1 presents examples of behavioral terms relevant to the various phenomena suggested by SA definitions. We selectively focus on the three levels of SA because the behavioral phenomena involved at each level suggest stimuli to manipulate for the purposes of predicting and controlling behavior.

Perception

Within the SA framework, perception refers to sensing stimuli in the environment. Although a stimulus can be a characteristic feature of an object, in SA, perception includes sensing the object as a stimulus as well as the relevant characteristics of that object (such as the formations in a mountain range; Endsley, 1995a). Within behavior analysis, perception can be determined only when a response occurs following the presentation of a stimulus

(Schlinger, 1995). For example, a pilot flying through fog would not be said to perceive a mountain range on a radar display unless he acts in accordance with the implications of the graphic display. It should be noted that the action could be as subtle as orienting to the stimulus (in this case the display). This concept (i.e., *stimulus control*) entails the consistent effects of a stimulus on a response (Catania, 2013). In that context, if the pilot consistently responds to the display, then the responses would be under the control of that stimulus.

In a dynamic situation, it is unlikely that just one stimulus controls a given response. The concept of *conditional discrimination* is relevant, as it refers to the operant control of a stimulus in the context of other stimuli (Catania, 2013). That is, a given stimulus, in the presence of other stimuli, signals the availability of reinforcement contingent on a particular response. In the flying through fog example, a buzzing noise and flashing lights from the plane's ground proximity warning system, combined with a representation of a mountain on a computer-generated display, occasion a class of behavior that functions to avoid the mountain. When paired with clear skies and no land in sight, the same light and sound occasion classes of behavior with different results: the faulty sensor is shut off and backup systems are brought online. Though the relations between stimuli and responses in a dynamic task are infinitely complex, the stimulus control and conditional discrimination concepts are useful to distill the phenomena into critical features and functions of stimuli and responses. Doing so creates objective and functional units of behavior that can be analyzed and modified for the purposes of prediction and control.

Perception is not considered behavior in traditional SA. In a behavioral account, behavior is comprised of all the overt and covert activities of the person; so, perceiving becomes an action tied to a stream of prior events. Thus, the analysis of perception can expand to include the previous behaviors that produce additional stimuli relevant to the task. That is, it can be useful to identify the class of behaviors that increase the performer's contact with stimuli that have been correlated with current operating contingencies in a task. In the experimental analysis of behavior, this is called an *observing response*, and it refers to a class of behavior that produces stimuli correlated with the components of a compound schedule of reinforcement (Wyckoff, 1952). The nurse who checks for drug allergies (the observing response) before injecting antibiotics into a septic patient is more likely to be successful than a nurse who does not. Another example is when a driver looks at a cell phone, misses discriminative stimuli around the vehicle, and therefore misses opportunities to make responses critical to safe driving (e.g., increasing distance when following a driver that is swerving). These missed opportunities affect a performer's ability to engage in other classes of behavior involved in subsequent comprehension and projection (Kass, Cole, & Stanny, 2007). In a behavioral account of perception, the responses that produce discriminative stimuli can be just as important as the discriminative effects of the stimuli on behavior. Thus, Level 1 should expand to include the responses that enable contact with relevant task stimuli.

A *learning channel* is an additional concept that is useful in the analysis of perception (Haughton, 1980; Kubina & Yurich, 2012; Lindsley, 1998). The learning channel directs an observer to identify the specific stimuli and corresponding response by analyzing the stimulus and response modalities of the action being performed. In the case of the flight display and ground proximity buzzer, the discriminative properties are visual and auditory; therefore, the first part of the learning channel would be *see-hear*. If the action required in the presence of these stimuli is to pull back on flight controls, then the second component of the learning channel is *touch*. The entire learning channel with respect to the stimuli is *see-hear-touch*, with the *see-hear* representing the perception phase of SA. Or, the response may be to tell the copilot to pull back on the control column, which involves the *see-hear-say channel*. Using the learning channel framework, an analysis of how an individual reacts in a given situation focuses on the observable stimulus-response interactions. Such an analysis adds to the SA model by incorporating the functional analytic unit (i.e., the operating contingencies that make "perception" work in an applied environment). Additionally, application of the learning channels concept removes the need to speculate about the effects of the mental agents on task outcomes and focuses the description of perception on observable stimuli and events.

Comprehension

In the traditional SA model, comprehension is understood as the actions of covert mental structures and processes. That is, once a performer perceives stimuli, the mental mechanisms process the "meaning" of the stimuli in relation to the present task and task goals (both of which are stored in memory). A behavioral model of comprehension does not require a reference to hypothetical mental agents. A behavior analytic interpretation can account for the process by which verbal behavior comes under the control of stimuli and complex stimulus-stimulus relations in a task. Comprehension behavior seems to involve tacting the features and functions of task stimuli and the ways in which they relate to other task stimuli and events. For example, a skilled pump operator at a community water supply can fluently state the effects of moving levers and pressing buttons at a control panel and how balancing multiple pump functions relates to ongoing water distribution when demand rises and some pumps malfunction. It is possible that these skills enable the operator to be relatively more effective in a problem-solving situation compared to a novice pump operator who cannot state the relationships. Of course, this hypothesis requires a systematic assessment of the role of verbal behavior in performance. In this example, the analysis is focused on categorizing verbal behavior based on putative function in pump operations.

Developing a behavioral assessment strategy also entails an analysis of how comprehension responses function to alter subsequent responding (i.e., response-response relations) and ultimately affect task performance. Topographically, the contents of a comprehension response include current sources of stimulation and how they are related to other stimuli and the task at hand. Functionally, the comprehension response occasions other responses critical to achieving task outcomes. That is, after one emits discriminated responses in the presence of relevant task stimuli, a subsequent verbal statement is emitted and this sets the occasion for concurrent or subsequent responding. With respect to learning channels, comprehension is the *say* portion of *hear-see-say*. Previous observing and orienting responses, one's history with similar events, and competing stimuli influence the accuracy, speed, and complexity of a comprehension response.

Analyzing the function of a comprehension response aligns with Skinner's (1957) perspective on the concept of "meaning": Comprehension responses are not determined by structure alone but by their functional relations in a chain of behaviors and the ultimate outcome(s). A behavior analysis is poised to investigate whether comprehension responses evoke other responses that have produced reinforcing stimuli. A more molecular analysis involves investigating whether the accuracy and speed of comprehension responses are predictive of the accuracy and speed of projection responses during a task. Returning to the example of the pump operator, a red light above a valve number on a display panel signals that a valve has failed, and this should occasion responses that function to divert water flow to secondary pathways. An untrained operator sees the light but lacks the training to tact the state of the valve and produce behavior to continue pumping. Alternatively, the trained operator makes a "comprehension" statement that the valve has malfunctioned and other pathways are available. In this situation, the comprehension response increases the probability of successful pump function and decreases the probability of a shutdown.

Expanding the concept of the comprehension response could involve an analysis of the contingencies between the comprehension responses and the behavior of others in a group problem-solving scenario. This entails analyzing a chain of interlocked behaviors by members of a work team and the alteration in the probability of reinforcement enabled by comprehension responses. Many behaviors examined under the SA umbrella involve coordination of multiple workers, so an analysis of the verbal behavior that enables group problem solving is critical to adaptive behavior and task outcomes.

From a behavioral perspective, an intervening comprehension response is not always necessary to engage in subsequent behavior that affects task performance. Upon seeing a light that indicates valve failure, an operator may simply manipulate controls without emitting verbal behavior. In fact, many masterful performances appear to occur "automatically" in which performers may barely acknowledge particular components of their behavior (Bloom, 1986; Dougherty & Johnston, 1996). By incorporating a functional account of comprehension responses in complex tasks, behavior analysis adds flexibility in identifying critical behaviors, verbal or otherwise, that affect the total task performance.

Projection

Projecting, or predicting, is behaving with respect to the current functions of stimulus events based on a history of contact with those stimulus functions (Skinner, 1974). For example, one could argue that a rat predicts the delivery of a food pellet as it presses a lever because the lever has produced pellets in the past. In this sense, prediction is simply behaving in the present, as a part of present stimulus conditions (L. J. Hayes, 2001), which have their effect due to experiences in the past. It is important to note that predicting does not mean that behavior is affected by future events. It is impossible for future events to affect present events because they have not happened, and there is no event that has not happened that can affect the present stimulus functions (L. J. Hayes, 1992; Skinner, 1974). Therefore, no portion of our present behavior is a function of future circumstances. Instead, behavior is a function of the contextual conditions of the psychological present.

As humans, we exhibit verbal behavior that enables behaving with respect to a vast network of interrelated stimulus functions, greatly expanding our range of prediction. An individual can respond to the relation between stimulus functions (S. Hayes, Barnes-Holmes, & Roche, 2001), such as a truck driver behaving with respect to speed, fuel level, traffic patterns, and road conditions. If a tire suddenly blows out as a semi-truck driver is rapidly approaching a curve on a mountain road, the driver can verbally construct possible scenarios describing what may happen if he continues on or changes what he is doing, even if he has never encountered such a situation. It could be said that the driver is predicting a safe stop by either downshifting or braking depending on the conditional stimuli. That is, the driver's behavior is influenced by the multitude of stimulus functions that combine in new ways and produce effective action in the novel situation. It might be tempting to argue that predicting means acting so that things will turn out a certain way, or in hopes that they will turn out a certain way. However, a caveat to predicting as "action toward the future" is that the ability to predict is enabled by a history of similar response and stimulus functions in the past (e.g., reinforcement). The more parsimonious historical analysis renders the teleological hypothesis unnecessary.

Novel predictive behavior under the control of rules might be dismissed by some as solely an instance of rule following. However, explaining the behavior as rule following does not adequately describe how that behavior occurs. In the example involving the semi-truck driver, activating the brakes might be in a frame of coordination with the verbal stimulus "pull the emergency brakes." This is part of a network of relations: if stimulus event A (blown tire) in conjunction with stimulus events B (dangerous road conditions) and C (no cars immediately behind the truck), then stimulus D (emergency brake) has a new function of evoking pulling behavior. Lever pulling would be a verbal event because it is under the control of a verbal stimulus (Skinner, 1957). A behavioral approach that examines the development and generalization of verbal behavior classes involves assessing rule-stimuli relations and testing for generalization to novel events. Simply repeating a series of rules may not be a sufficient repertoire for effective performance under novel conditions; in fact, it may establish a repertoire that invites insensitivity to task relevant stimuli (Wulfert, Greenway, Farkas, Hayes, & Dougher, 1994).

As with comprehension, a functional analysis of prediction involves assessing the connection of predictive statements to task performance. A semi-truck driver need not emit a predictive statement when a tire blows out. In some cases, the driver simply needs to pull off to the side of the road to maintain safety. However, a predictive statement may be beneficial for the same driver to plan lane changes at an upcoming highway interchange during rush hour. A research agenda that links verbal behavior to prevalent fluctuating contingencies and task outcomes can help realize the importance of predictive statements and identify critical differences between experts and novices on tasks.

In essence, the three levels of SA analyzed from a behavioral perspective expands the conceptualization of phenomena involved in dynamic tasks and lends itself to the investigation of core behavioral units predictive of other behavior and task outcomes. This analysis directs researchers and practitioners to the effective arrangement of environmental stimuli and events, whether it is through manipulating antecedents (e.g., the labeling of controls on a panel) or establishing a history of behavior-environment relations (e.g., training). Though there are obvious discrepancies between the mental constructs in SA and behavioral principles, partitioning an infinitely complex continuum of events into the perception, comprehension, and projection sequence enables a practical analysis. A starting point for behavior analysts is to identify the ways in which SA assessment tools are used. SA assessments can be effectively employed by behavior analysts to predict and control behavior because the assessments reveal relationships between environmental stimuli and responses in work situations.

SA assessment tools

A number of specific SA assessment tools exist. These tools are often used in a similar fashion as protocol analyses (Austin & Delaney, 1998; Ericsson & Simon, 1993; S. Hayes, 1986; S. Hayes, White, & Bisset, 1998) and talk-aloud problem solving (TAPs; Berardi-Coletta, Buyer, Dominowski, & Rellinger, 1995; Pate, Wardlow, & Johnson, 2004; Whimbey & Lochhead, 1999). These methods require participants to narrate the task, so researchers can identify putative environmental variables that control behavior. The methods also identify rules and other verbal behavior that lead to ineffective problem solving. Incorrect and delayed answers on SA tests direct researchers toward faulty stimulus control due to equipment design, poor problem-solving strategies, and ineffective training methods. In addition, expert performers tested with SA measures can reveal a rich source of variables to be instituted in training novices and designing equipment for optimal performance. SA assessments inform the design of testable solutions for instructional designers and equipment engineers because the assessments provide a link between antecedent stimuli, verbal behavior, and task outcomes. That is, SA assessments are useful for testing the effects of both antecedent and consequence-based interventions.

The Situation Awareness Global Assessment Technique (SAGAT) pioneered by Endsley (1987, 1988b, 1995b) is a popular measure in SA research (Durso & Gronlund, 1999). The initial phase of SAGAT is to determine the stimuli and behaviors that are critical to task performance. To achieve this, experimenters interview experts on the given task (e.g., flying an airplane, operating a piece of heavy equipment). A new performer then engages in a simulated task that incorporates the critical stimuli. There are interruptions throughout the task in which the activity is halted (or the computer interface turns completely black), and the performer is queried on the status and location of the relevant task stimuli (perception), the meaning of these stimuli (comprehension), and predictions of what is likely to happen if the simulation continues (projection). Accuracy of response (i.e., percent correct) is the dependent variable that determines the level of SA. In essence, SAGAT is a process for developing a customized assessment and a procedure to implement the assessment rather than a particular user interface, task, or a set of questions. This inherent flexibility allows the process to cover a broad range of tasks.

A majority of the early studies utilizing SAGAT focused on aviation (e.g., Endsley, 1988a; Endsley, Mogford, Allendoerfer, Snyder, & Stein, 1997). However, recent studies have verified the positive correlation between SAGAT performance and task performance on diverse topics such as obstetrical team crisis response (Morgan et al., 2015), petrochemical control room operations (Ikuma, Harvey, Taylor, & Handal, 2014), driving distractors (Kass et al., 2007; Ma & Kaber, 2005) and military planning (Salmon et al., 2009).

Another SA assessment is the Situation Present Assessment Method (SPAM; Durso & Dattel, 2004). SPAM is an attempt to advance upon SAGAT by incorporating an additional stage of query to ascertain "workload." This is achieved by including time as a dimension of verbal responses. The procedures for SPAM are similar to SAGAT in that a performer participates in a simulated task. During the task, the performer is given a warning signal that a query is imminent, and the participant chooses when to answer the query. The delay from the onset of the warning signal to the choice to answer the query is considered a measure of workload. Once the query begins, the task is paused and a series of questions similar to SAGAT are presented. Accuracy and speed of query responses are measured and assumed to represent SA. There is initial evidence to validate SPAM as a useful measurement tool. For example, answers to SPAM predicted performance on an air traffic control (Durso et al., 1998) and predicted performance relative to screen display configuration in a simulated submarine-tracking task (Loft & Morrell, 2013).

Directly assessing performance on a specific task is another approach to SA measurement, though it is less common. Andre, Wickens, Moorman, and Boschelli (1991) used a low-fidelity flight simulator to assess the response times of "flight-naïve" participants. The goal was to identify the best arrangement of components on a computerized flight display by testing flight perspective (two-dimensional vs. three dimensional), frame of reference (inside-out vs. outside-in), and visual momentum (monochrome vs. color). The flight task consisted of flying the simulated aircraft through a series of on-screen rings. At random points during the task, the screen would briefly go blank and return with the plane at a different bank, pitch, altitude, and heading. The dependent variable, and measure of SA, was the amount of time it took the pilot to adjust the plane back to the original heading. Results indicated that time to completion was significantly shorter with outside-in reference, color, and two-dimensional displays, respectively. The assessment used by Andre et al. (1991) is similar to SAGAT and SPAM in that overt performance is considered indicative of underlying SA; however, the Andre et al. (1991) study differs because the ultimate performance of interest (response accuracy and duration to an unexpected event) was measured while verbal responses related to prediction, comprehension, and projection were not.

Behavioral research and SA

For behavior analysts working in organizations, the technologies developed through SA show promise for designing better human-machine interfaces, systemization of training, and assessment of mastery; all of these areas are addressed in the OBM literature. For behavior analysts interested in applying

SA technologies, the following is a discussion of considerations in using simulations, the relevance of fluency research to SA, and opportunities for research in OBM. The discussion is primarily directed toward behavior analysts who conduct applied research and see the prospect of SA methodologies to advance behavioral assessment and intervention.

Considerations for simulations

On the surface, simulating dynamic tasks is a potential barrier to conducting research on SA phenomena. In traditional SA research, high- and low-fidelity computer programs are often used to simulate tasks such as flying airplanes and tracking submarines. The time and money dedicated to developing these programs can be substantial; however, trends suggest the increasing ease of programming for non-computer science professionals (Potter, Roy, & Bianchi, 2014). As cohorts of behavior analysts exit graduate programs with basic computer programming competence, developing basic simulators and learning programs is within reach (Escobar, 2014; Potter, Roy, & Bianchi, 2014; Twyman, 2014).

Though software is attractive because of its convenience and low safety risk, it is not always necessary to produce realistic simulations. For example, researchers investigating emergency medical response to trauma events have used standard work equipment and actor-patients to create simulations (e.g., Cooper et al., 2010). Simulations can be further enhanced by incorporating low-cost technologies such as GoPro™ cameras to track responses. Furthermore, mobile devices can deliver SA queries on a given schedule, or the query can be sent from another mobile device controlled by a researcher. As the operation of equipment is increasingly controlled via computers with touchscreens, software such as Camtasia (Techsmith, 2016) can record the participant's entire interaction with the screen. This enables response tracking without the need for a sophisticated program to interact with the equipment software. Such solutions are likely to emerge in industries that are automating operations (e.g., oil rigs, power plants), and the workers shift from manual labor to operating equipment remotely.

Fluency research and SA

Methods in precision teaching and the concept of behavioral fluency can assist in the development of SA analyses and training methodologies. Particularly, the fluency of functional verbal behavior classes is pertinent to SA and task performance. Assessment in precision teaching uses response rate and latency to identify whether classes of behavior (a) retain after periods of no practice, (b) can endure for long periods without a reduction in rate or quality, (c) can maintain rate despite distracters, (d) can be functionally applied across a variety of settings and stimuli, and (e) are available to be combined with other classes of behavior (Binder, 1996; Kubina & Yurich, 2012). These measures and outcomes align with the goals of SA researchers in designing better equipment and training strategies to foster expedient and effective performance in novel conditions (Endsley & Jones, 2011; Roberts et al., 2016).

Traditionally, fluency research has been conducted in K–12 education; however, in a general sense the topics addressed are similar to those in traditional SA research. Problem-solving, reasoning, and predicting are examples of complex repertoires measured in precision teaching and enhanced by rate-building procedures on relatively less complex behavior classes (i.e., tool skills; Johnson & Street, 2004). As tool skills increase in rate, a learner is exposed to novel problems and contingencies to promote the emergence of a new, more complex behavior class (Johnson & Layng, 1996). By incorporating this detailed approach of behavior into SA assessments and interventions, predictions of task outcomes and the identification of effective intervention components can be refined. For example, the latency for a performer to make an appropriate observing response to a control panel item or the rate at which one can state relations between task stimuli are potentially more sensitive measures than accuracy alone because they incorporate time as a response dimension (Johnston & Pennypacker, 2008).

The prediction and control of untrained problem solving in high-stress scenarios could be enhanced by identifying core behavior units, assessing baselines, and training the behaviors to frequency aims. This approach encourages economical analysis and intervention because training on all potential scenarios and response classes is not necessary; rather, the focus would be on assessing and training members of a response class that encourage the novel combination of other response classes under novel contingencies. Using this strategy in SA and OBM approximates a model of "generative instruction" that has traction in 21st century training for both children and adults (Johnson, 2015).

Research opportunities in OBM

Verbal behavior and dynamic tasks

Perhaps the most needed line of research in approaching SA from a behavioral perspective is identifying the function of behaviors traditionally categorized as perception, comprehension, and projection. The molecular approach seeks to identify sources of variation for these elemental behaviors and how these responses occasion other responses or produce task-relevant outcomes. For example, the "call and response" type communication used by emergency medical staff in a crisis scenario can affect the accuracy and speed of responses on perception, comprehension, and projection questions. A molar approach seeks to identify how certain responses affect total-task performance. For example, call and response communication can affect the total amount of time and accuracy to diagnosis and medical intervention (Outcome 1), and this can significantly impact the probability of patient recovery (Outcome 2). Both molar and molecular analyses are likely useful in pinpointing classes of behavior common in dynamic tasks.

Because SA measures can pinpoint verbal processes, various verbal behavior taxonomies can be applied in analysis of performance in dynamic situations. For example, relational frame theory (S. Hayes et al., 2001) could be compared to a Skinnerian account of verbal behavior in an attempt to evaluate the utility of constructs for analysis and intervention in emergency medical response. This would be a unique contribution, and it is relevant to the call for incorporating alternative accounts of language in OBM (S. Hayes, Bunting, Herbst, Bond, & Barnes-Holmes, 2006).

A behavioral perspective advances the SA approach to perception by prescribing an investigatory framework for the role of verbal behavior in establishing stimulus control and, by implication, how stimulus control breaks down in a dynamic situation (Alavosius, Houmanfar, & Rodriquez, 2005). In doing so, rules, instructions, and training for high-risk jobs may be analyzed for their efficacy in relation to performers with a wide range of learning histories. This is relevant to contexts where training to mastery is unavailable due to budget and time restriction. It is possible that training adherence to specific, tactical rules will occasion responses that are insensitive to dynamic and changing contingencies (Wulfert et al., 1994). It is also possible that more strategic rules, such as "during an emergency situation, stop, scan, and think about what you're going to do," occasion adaptive responding that increases the probability of reinforcement under the changing conditions. A method that identifies critical classes of verbal behavior in a real or simulated job task would certainly be a worthy contribution to the OBM and SA literatures.

Simulator fidelity and transfer of training

The level of fidelity between computerized-simulators, in-vivo simulations, and actual performance sites is another area needing research. It may be that expensive high-fidelity simulators are not needed to teach SA in all settings. Simple, low-cost simulators for research and training may be functionally elegant and capture the critical behavior-context relations from which other responses emerge. This research line could focus on an assessment methodology for determining the minimal simulation components that allow for maximum transfer to on-the-job performance and novel problem solving. Additionally, this allows organizations access to functional simulations when high-fidelity equipment is cost prohibitive.

The taxonomy suggested by Stokes and Baer (1977) offers a starting point to identify techniques of generalization used with simulators. These techniques are

(a) train and hope, (b) sequential modification, (c) introduce to natural maintaining contingencies, (d) train sufficient exemplars, (e) train loosely, (f) use indiscriminable contingencies, and (g) programing common stimuli. In a given task, for example, it may be more effective to train with multiple exemplars instead of focusing on creating hyper-realistic 3D simulations that encourage the learner to generate specific rules governing behavior. This is particularly relevant if the goal is to produce novel problem solving rather than adherence to a fixed set of procedures. Stokes's and Baer's (1977) taxonomy and generalization programming procedures (Stokes & Osnes, 1989) enable a systematic approach to identifying the technologies of generalization that most support the transfer of skills to dynamic work environments.

Conclusion

This paper discussed the predominant model of SA in human factors research, a behavioral interpretation of the events addressed by this model, common SA assessment tools, and the relevance of SA to behavioral research and application. Looking beyond the theory and hypothetical mental mechanisms in SA reveals a method to systematically partition an infinitely complex sequence of events into manageable analytical units. Traditional SA research has laid some of the groundwork for behavior analysts to bring an inductive approach to describing, assessing, predicting, and controlling performance in complex environments. A behavioral interpretation adds depth and breadth to the SA analysis by identifying additional sources of behavioral variability to address complex stimulus-stimulus, stimulus-response, and response-response relations.

SA and dynamic tasks present a wealth of research questions such as the link between classes of verbal behavior and problem solving, fluency and component-composite skill relations in work tasks, and the relationship between simulator fidelity and the generalization of behavior classes. Contemporary SA research is productive, and the constructs underlying this work continue to generate effective solutions in training, equipment design, and more. Behavior analysts can use current SA research methods in developing replicable behavioral assessments and achieve the acceptance SA has won in the marketplace of behavioral technologies.

References

Alavosius, M. P., Houmanfar, R., & Rodriquez, N. J. (2005). Unity of purpose/Unity of effort: Private-sector preparedness in times of terror. *Disaster Prevention and Management, November, 14*, 666–680. doi:10.1108/09653560510634098

Andre, A. D., Wickens, C. D., Moorman, L., & Boschelli, M. M. (1991). Display formatting techniques for improving situation awareness in the aircraft cockpit. *The International Journal of Aviation Psychology, 1*, 205–218. doi:10.1207/s15327108ijap0103_2

Austin, J., & Delaney, P. F. (1998). Protocol analysis as a tool for behavior analysis. *The Analysis of Verbal Behavior, 15*, 41–56.

Baumgartner, N., Gottesheim, W., Mitsch, S., Retschitzegger, W., & Schwinger, W. (2010). BeAware! – Situation awareness, the ontology-driven way. *Data & Knowledge Engineering, 69*(11), 1181–1193. doi:10.1016/j.datak.2010.07.008

Behzad, B., & Mann, D. D. (2014). Automation and the situation awareness of drivers in agricultural semi-autonomous vehicles. *Biosystems Engineering, 124*, 8–15. doi:10.1016/j.biosystemseng.2014.06.002

Berardi-Coletta, B., Buyer, L. S., Dominowski, R. L., & Rellinger, E. R. (1995). Metacognition and problem-solving: A process-oriented approach. *Journal of Experimental Psychology: Learning, Memory and Cognition, 21*, 205–223. doi:10.1037/0278-7393.21.1.205

Billings, C. E. (1995). Situation awareness measurement and analysis: A commentary. In D. J. Garland, & M. R. Endsley (Eds.), *Experimental analysis and measurement of Situation Awareness*. Daytona Beach, FL: Embry-Riddle Aeronautical University Press.

Binder, C. (1996). Behavioral fluency: Evolution of a new paradigm. *The Behavior Analyst, 19*, 163–197.

Bloom, B. S. (1986). Automaticity: The hands and feet of genius. *Educational Leadership, 43* (5), 70–77.

Carol, L. A. (1992). Desperately seeking SA. *TAC Attack, 32*(3), 5–6.

Catania, A. C. (2013). *Learning* (5th ed.). Cornwall-on-Hudson, NY: Sloan Publishing.

Cooper, S., Kinsman, L., Buykx, P., McConnell-Henry, T., Endacot, R., & Scholes, J. (2010). Managing the deteriorating patient in a simulated environment: Nursing students' knowledge, skill and situation awareness. *Journal of Clinical Nursing, 19*(15–16), 2309–2318. doi:10.1111/j.1365-2702.2009.03164.x

Dougherty, K. M., & Johnston, J. M. (1996). Overlearning, fluency, and automaticity. *The Behavior Analyst, 19*, 289–292.

Durso, F. T., & Dattel, A. (2004). SPAM: The real-time assessment of SA. In S. Banbury, & S. Trembley (Eds.), *A cognitive approach to situation awareness: Theory, measures and application* (pp. 137–154). New York, NY: Aldershot.

Durso, F. T., & Gronlund, S. D. (1999). Situation awareness. In F. T. Durso, R. S. Nickerson, R. W. Schvanveldt, S. T. Dumais, D. S. Lindsay, & M. T. H. Chi (Eds.), *Handbook of applied cognition* (pp. 283–314). Hoboken, NJ: John Wiley & Sons Ltd.

Durso, F. T., Hackworth, C. A., Truitt, T. R., Crutchfield, J., Nikolic, D., & Manning, C. A. (1998). Situation awareness as a predictor of performance for en route air traffic controllers. *Air Traffic Control Quarterly, 6*, 1–20. doi:10.1207/s15327108ijapo2_6

Endsley, M. R. (1987). *SAGAT: A methodology for the measurement of situation awareness* (NOR DOC 87-83). Hawthorne, CA: Northrop Corp.

Endsley, M. R. (1988a). Design and evaluation for situation awareness enhancement. In *Proceedings of the human factors society 32nd annual meeting* (pp. 97–101). Santa Monica, CA: Human Factors and Ergonomics Society.

Endsley, M. R. (1988b). Situation awareness global assessment technique (SAGAT). In *Proceedings of the national aerospace and electronics conference (NAECON)* (pp. 789–795). New York, NY: IEEE.

Endsley, M. R. (1995a). Toward a theory of situation awareness in dynamic systems. *Human Factors, 37*, 32–64. doi:10.1518/001872095779049543

Endsley, M. R. (1995b). Measurement of situation awareness in dynamic systems. *Human Factors, 37*, 65–84. doi:10.1518/001872095779049499

Endsley, M. R., & Jones, D. G. (2011). *Designing for situation awareness: An approach for user-centered design*. Boca Raton, FL: CRC Press.

Endsley, M. R., Mogford, R., Allendoerfer, K., Snyder, M. D., & Stein, E. S. (1997). Effect of free flight conditions on controller performance, workload and situation awareness: A preliminary investigation of changes in locus of control using existing technology (DOT/FAA/CT-TN 97/12). Atlantic City, NJ: Federal Aviation Administration William J. Hughes Technical Center.

Ericsson, K. A., & Simon, H. A. (1993). *Protocol analysis: Verbal reports as data* (2nd ed.). Cambridge, MA: MIT Press.

Escobar, R. (2014). From relays to microcontrollers: The adoption of new technology in operant research. *The Mexican Journal of Behavior Analysis, 40*(2), 127–153.

Haines, R. F., & Flateau, C. (1992). *Night flying.* Blue Ridge Summit, PA: TAB Books.

Haughton, E. C. (1980). Practicing practices: Learning by activity. *Journal of Precision Teaching, 1*(3), 3–20.

Hayes, L. J. (1992). The psychological present. *The Behavior Analyst, 15,* 139–146.

Hayes, L. J. (2001). Finding our place in the constructed future. In L. J. Hayes, J. A. Austin, R. Houmanfar, & M. C. Clayton (Eds.),.*Organizational Change.* Reno, NV: Context Press.

Hayes, S. C. (1986). The case of the silent dog - verbal reports and the analysis of rules: A review of Ericsson and Simon's Protocol Analysis: Verbal Reports as Data. *Journal of the Experimental Analysis of Behavior, 45,* 351–363. doi:10.1901/jeab.1986.45-351

Hayes, S. C., Barnes-Holmes, D., & Roche, B. (Eds.). (2001). *Relational frame theory: A post-Skinnerian account of human language and cognition.* New York, NY: Kluwer Academic/Plenum Press.

Hayes, S. C., Bunting, K., Herbst, S., Bond, F. W., & Barnes-Holmes, D. (2006). Expanding the scope of organizational behavior management: Relational frame theory and the experimental analysis of complex behavior. *Journal of Organizational Behavior Management, 26* (1–2), 1–23. doi:10.1300/J075v26n01_01

Hayes, S. C., White, D., & Bissett, R. T. (1998). Protocol analysis and the "silent dog" method of analyzing the impact of self-generated rules. *The Analysis of Verbal Behavior, 15,* 57–63.

Ikuma, L. H., Harvey, C., Taylor, C. F., & Handal, C. (2014). A guide for assessing control room operator performance using speed and accuracy, perceived workload, situation awareness, and eye tracking. *Journal of Loss Prevention in the Process Industries, 32,* 454–465. doi:10.1016/j.jlp.2014.11.001

Johnson, K. (2015). Behavioral education in the 21st century. *Journal of Organizational Behavior Management, 35*(1–2), 135–150. doi:10.1080/01608061.2015.1036152

Johnson, K. J., & Layng, T. V. J. (1996). On terms and procedures: Fluency. *The Behavior Analyst, 19,* 281–288.

Johnson, K. J., & Street, E. M. (2004). *The Morningside model of generative instruction: What it means to leave no child behind.* Concord, MA: Cambridge Center for Behavioral Studies.

Johnston, J. M., & Pennypacker, H. S. (2008). *Strategies and tactics of behavioral research* (3rd ed.). New York, NY: Routledge.

Kass, S. J., Cole, K. S., & Stanny, C. J. (2007). Effects of distraction and experience on situation awareness and simulated driving. *Transportation Research Part F: Traffic Psychology and Behaviour, 10,* 321–329. doi:10.1016/j.trf.2006.12.002

Kubina, R., & Yurich, K. (2012). Behavioral fluency. In *The precision teaching book* (pp. 317–347). Lemont, PA: Greatness Achieved Publishing Company.

Lindsley, O. R. (1998). Learning channels next: Let's go! *Journal of Precision Teaching and Celeration, 15,* 2–4.

Loft, S., & Morrell, D. B. (2013). Using the situation present assessment method to measure situation awareness in simulated submarine track management. *International Journal of Human Factors and Ergonomics, 2,* 33–48. doi:10.1504/ijhfe.2013.055975

Ma, R., & Kaber, D. B. (2005). Situation awareness and workload in driving while using adaptive cruise control and a cell phone. *International Journal of Industrial Economics, 35* (10), 939–953. doi:10.1016/j.ergon.2005.04.002

Morgan, P., Tregunno, D., Brydges, R., Pittini, R., Tarshis, J., Kurrek, M., ... Ryzynski, A. (2015). Using a situational awareness global assessment technique for interprofessional obstetrical team training with high fidelity simulation. *Journal of Interprofessional Care, 29*, 13–19. doi:10.3109/13561820.2014.936371

Omnasch, L., Wickens, C. D., Li, H., & Manzey, D. (2014). Human performance consequences of stages and levels of automation: An integrated meta-analysis. *Human Factors, 56*(3), 476–488. doi:10.1177/0018720813501549

Pate, M. L., Wardlow, G. W., & Johnson, D. M. (2004). Effects of thinking aloud pair problem solving on the troubleshooting performance of undergraduate agriculture students in a power technology course. *Journal of Agricultural Education, 45*(4), 1–11. doi:10.5032/jae.2004.04001

Potter, W., Roy, R., & Bianchi, S. (2014). Computer programming for research and application: Livecode development environment. *Mexican Journal of Behavior Analysis, 40*(2), 154–191.

Roberts, R., Flin, R., & Cleland, J. (2015). "Everything was fine"*: An analysis of the drill crew's situation awareness on Deepwater Horizon. *Journal of Loss Prevention in the Process Industries, 38*, 87–100. doi:10.1016/j.jlp.2015.08.008

Roberts, R., Flin, R., & Cleland, J. (2016). How to recognise a kick: A cognitive task analysis of drillers' situation awareness during well operations. *Journal of Loss Prevention in the Process Industries, 43*, 503–513. Retrieved July 9, 2016. doi:10.1016/j.jlp.2016.07.003

Salmon, P. M., Stanton, N. A., Walker, G. H., Jenkins, D., Ladva, D., Rafferty, L., & Young, M. (2009). Measuring Situation Awareness in complex systems: Comparison of measures study. *International Journal of Industrial Economics, 39*(3), 490–500. doi:10.1016/j.ergon.2008.10.010

Schlinger, H. D. (1995). *A behavior-analytic view of child development*. New York, NY: Plenum.

Skinner, B. F. (1953). *Science and human behavior*. New York, NY: Macmillan.

Skinner, B. F. (1957). *Verbal behavior*. New York, NY: Appleton-Century Crofts.

Skinner, B. F. (1969). *Contingencies of reinforcement: A theoretical analysis*. New York, NY: Appleton-Century-Crofts.

Skinner, B. F. (1974). *About behaviorism*. New York, NY: Random House, Inc.

Smith, K., & Hancock, P. A. (1995). Situation awareness is adaptive, externally directed consciousness. *Human Factors, 37*, 137–148. doi:10.1518/001872095779049444

Soliman, M., & Mathna, E. K. (2009). Metacognitive strategy training improves driving situation awareness. *Social Behavior and Personality: An International Journal, 37*(9), 1161–1170. doi:10.2224/sbp.2009.37.9.1161

Stevens-Adams, S., Cole, K., Haass, M., Warrender, C., Jeffers, R., Burnham, L., & Forsythe, C. (2015). Situation awareness and automation in the electric grid control room. *Procedia Manufacturing, 3*, 5277–5284. doi:10.1016/j.promfg.2015.07.609

Stokes, T. F., & Baer, D. M. (1977). An implicit technology of generalization. *Journal of Applied Behavior Analysis, 10*, 349–367. doi:10.1901/jaba.1977.10-349

Stokes, T. F., & Osnes, P. G. (1989). An operant pursuit of generalization. *Behavior Therapy, 20*, 337–355. doi:10.1016/S0005-7894(89)80054-1

Techsmith. (2016). *Camtasia (Version 2.10.5)*. Okemos, MI: TechSmith Corporation. Retrieved from http://www.Techsmith.com

Twyman, J. S. (2014). Envisioning education 3.0: The fusion of behavior analysis and learning science and technology. *The Mexican Journal of Behavior Analysis, 40*(2), 20–38.

van de Merwe, K., Oprins, E., Eriksson, F., & van der Platt, A. (2012). The influence of automation support on performance, workload, and situation awareness of air traffic controllers. *The International Journal of Aviation Psychology, 22*(2), 120–143. doi:10.1080/10508414.2012.663241

Whimbey, A., & Lochhead, J. (1999). *Problem solving and comprehension* (6th ed.). Hillsdale, NJ: Erlbaum.

Wickens, C. D. (1995). The tradeoff of design for routine and unexpected performance: Implications of situation awareness. In M. R. Endsley, & D. J. Garland (Eds.), *Situation awareness analysis and measurement* (pp. 190–202). Mahwah, NJ: Lawrence Earlbaum Associates, Inc.

Wickens, C. D. (2008). Situation awareness: Review of Mica Endsley's 1995 articles on situation awareness theory and measurement. *Human Factors, 50*(3), 397–403. doi:10.1518/001872008X288420

Wulfert, E., Greenway, D. E., Farkas, P., Hayes, S. C., & Dougher, S. C. (1994). Correlation between self-reported rigidity and rule-governed insensitivity to operant contingencies. *Journal of Applied Behavior Analysis, 27*(4), 659–671. doi:10.1901/jaba.1994.27-659

Wyckoff, L. B., Jr. (1952). The role of observing responses in discrimination learning: Part I. *Psychological Review, 59*, 431–442. doi:10.1037/h0053932

Assessing and Preventing Serious Incidents with Behavioral Science: Enhancing Heinrich's Triangle for the 21st Century

Terry McSween and Daniel J. Moran

ABSTRACT
The rate of occupational injuries has been declining annually, but the rate of decline for fatalities has not kept a similar pace. Behavior-based safety (BBS) contributes to reducing personal injuries, and can be applied to preventing serious incidents. To address serious injuries with greater confidence requires a change in perspective on the causes of fatalities and serious injuries. Heinrich's safety triangle helps describe the ratio between minor incidents and major incidents, but is not adequate in helping to *predict* serious incidents. Adding a special subset to the safety triangle can assist safety practitioners in predicting and influencing such events. Extending the triangle to include more foundational root causes, such as leadership shortcomings and system failures, will expand the scope of the behavior analysis, and including greater specificity about the precursors to serious incidents will help the precision of the behavior analysis. The implications of the expanded triangle for amplifying the effectiveness of BBS for reducing serious incidents are discussed.

Preventing serious incidents is a crucial area of focus for the safety profession. Industry leaders highlight the need for increased vigilance to the high and stable annual rate of fatalities and serious injuries (FSIs; Bogard, Ludwig, Staats, & Kretschmer, 2015; Krause & Murray, 2012; McSween, 2015), especially when compared to the relatively larger decline of the rate of other incidents. The workplace fatality rate in the United States *increased* from 3.3 deaths per 100,000 workers in 2013 to 3.4 deaths per 100,000 workers in 2014 (Bureau of Labor Statistics, 2016). The same reports indicated an increase in fatalities in construction, private mining, quarrying, oil and gas extraction, and roadway incidents. Manuele (2013) points out that "companies with outstanding records showing reductions in less-serious injuries may not have had similar reductions for serious injuries and fatalities" (p. 51). Past president of the American Society of Safety Engineers (ASSE) Terrie S. Norris pointed out the crux of the issue by saying,

> Despite the dedicated efforts of ASSE's members, employers, workers, the U.S. Occupational Safety and Health Administration (OSHA) and the National Institute for Occupational Safety and Health (NIOSH), the fact that fatalities are not significantly decreasing should be a call for action, not complacency.(American Society of Safety Engineers, 2011)

For over two decades, the annual rate of occupational injuries in the United States has been declining, but the rate of decline for fatalities has not kept a similar pace, especially in the last several years (see Figure 1). The two sets of data appear to have a positive correlation; however, the lower line showing fatalities per 100,000 employees does not demonstrate a similar, steeper downward slope as the nonfatal injuries. Working to prevent serious incidents goes beyond simply looking at fatalities and also includes focusing on lost workday cases (LWCs). McSween (2015) investigated the LWC rate and the total recordable injury rate (TRIR) in manufacturing and construction. In both areas, the TRIR showed a reasonable decline while the LWC rate remained relatively flatter in both industries (see Figures 2 and 3). This disparate trend requires the attention of organizational behavior management (OBM) and behavior-based safety (BBS) professionals, and the traditional behavioral approach should be optimized to focus on preventing serious incidents. In order to properly address this issue, this paper will question and modify the conventional approach to safety processes.

BBS has always encouraged organizations to change the environmental conditions, and not just the behaviors of workers, in order to improve safety

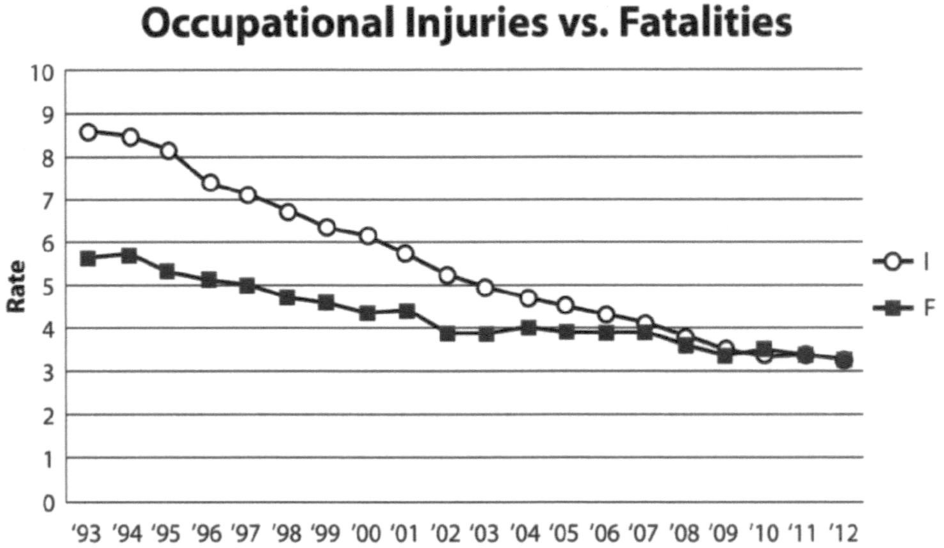

Figure 1. Fatalities per 100,000 employees compared to occupational injuries per 100 employees. Filled squares represent the annual fatalities per 100,000 employees in the United States. Open circles represent the annual injuries per 100 employees in the United States (United States Department of Labor, United States Department of Labor, Bureau of Labor Statistics, 2016).

Figure 2. Lost workday case (LWC) rate compared to total recordable injuries per 100 employees for U.S. manufacturing. Filled squares represent the LWCs and open circles represent the total recordable injuries per 100 employees (United States Department of Labor, United States Department of Labor, Bureau of Labor Statistics, 2016).

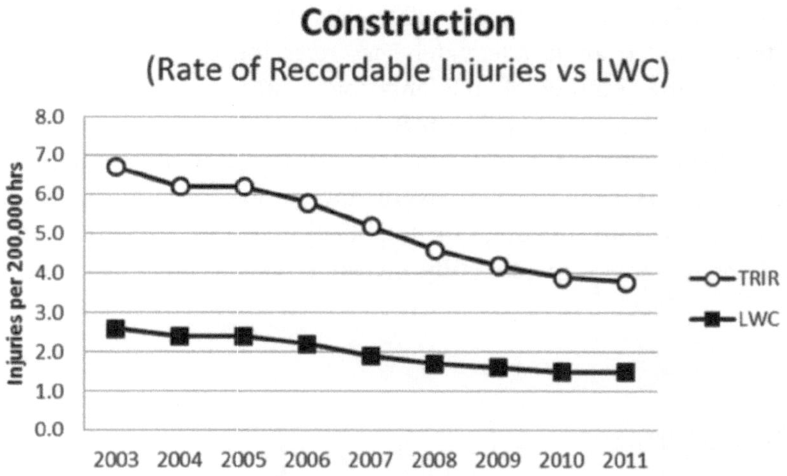

Figure 3. LWC rate compared to total recordable injuries (both per 100 employees) for U.S. construction companies. Filled squares represent the LWCs and open circles represent the total recordable injuries per 100 employees (United States Department of Labor, United States Department of Labor, Bureau of Labor Statistics, 2016).

outcomes. BBS can be more effective in preventing serious incidents if safety professionals embrace the concept that it is not just behavior, but *specific behaviors* that contribute to serious incidents, and that certain environmental *precursors* set up those specific behaviors. Those working to improve safety should not only focus on the specific behaviors, but on the causes of those behaviors. The causes of those behaviors are found in the design of the system, which may include physical hazards, leadership decisions, and other system failures that result in the rate of serious incidents and fatalities. This paper will first look at the traditional views of safety and behavior as represented by Heinrich's triangle, critique those views, and then provide a more focused approach to preventing serious incidents.

Heinrich's law: Popular but lacking predictive utility

Heinrich's Law (1931) and his "safety triangle" have become ubiquitous in the safety field, and the model implies that there is a standard ratio between the number of near misses, minor injuries, and major injuries at the worksite (see Figure 4). In the original design, Heinrich was implying that for every 300 near misses, there were 29 minor injuries, and 1 major injury. The triangle highlights the categories of incidents and the diminishing probability of the incident while going "up" the triangle vertically. Over the decades, safety professionals have refined the model, and often add other categories of incidents to the original triangle, suggesting that unsafe acts and conditions set the occasion for near misses, first-aid cases, recordable incidents, lost time incidents, and fatalities (McSween, 2003; see Figure 5).

The traditional triangle adequately *describes* the concept that there is a ratio between the number of incidents that could contribute to a possible fatality and an actual fatality (Martin & Black, 2015). Heinrich's work implies the probabilistic nature of catastrophic events: the greater the rate, duration, intensity, and perseverance of at-risk behaviors and at-risk conditions, the more likely a serious incident or fatality will occur. Because behavioral science aims to reduce incidents by investigating the environmental events

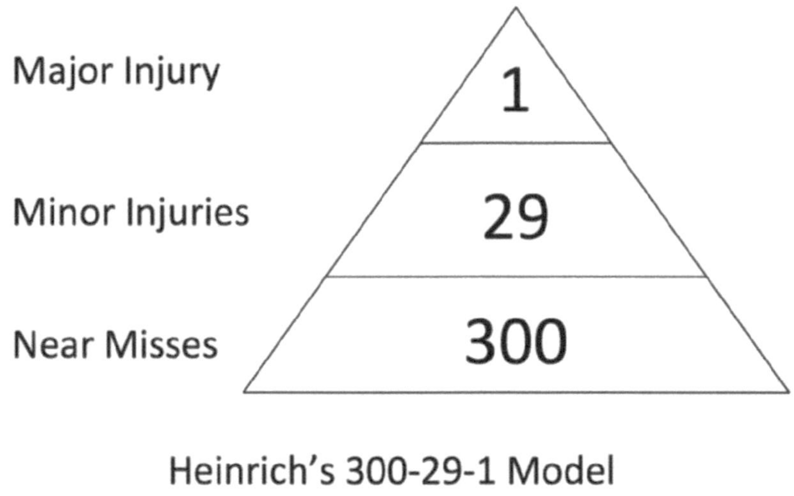

Figure 4. Heinrich's 300–29-1 model (Heinrich, 1931).

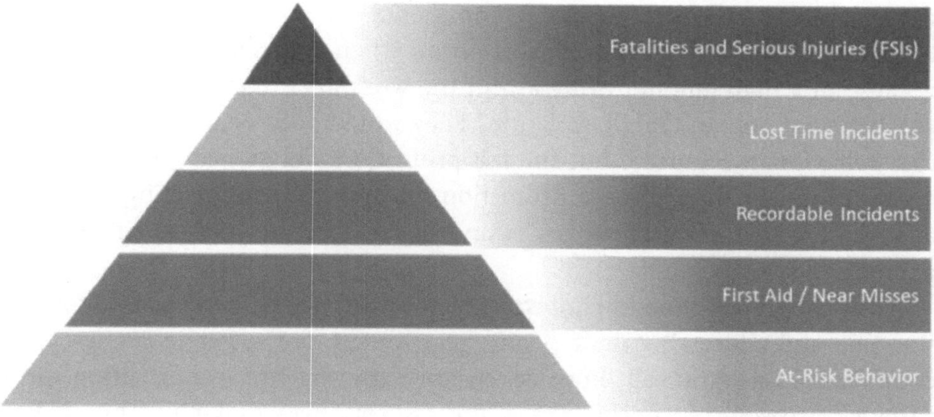

Figure 5. The traditional safety triangle expanded to include other factors (McSween, 2003).

that set the occasion for the unsafe actions and conditions, the triangle has heuristic value.

However, the triangle does not adequately *predict* incidents that lead to fatalities. OSHA and other governing bodies show that injury rates are poor predictors of FSIs (McSween, 2015). The aforementioned data trends showing minor injury rates continuing to drop while serious incident rates are plateauing suggest Heinrich's triangle lacks predictive utility. Krause and Murray (2012) suggest that the "absence of minor injuries is not predictive of an absence of future fatalities" (p. 2), and one dataset (RAND Corporation, 2007) suggests being "leery of drawing overarching conclusions about whether OSHA violations are likelier to contribute to deaths" (p. 132). Ultimately, the traditional triangle does not assist in projecting *which* of the unsafe acts or conditions will lead to a serious incident or fatality. In fact, many of Heinrich's ideas have been questioned (cf. Manuele, 2002). While the triangle is useful in conceptualizing the relationship between different types of injuries, the original ratios have little utility and may restrict our understanding of serious injuries.

Identifying precursors to FSIs using form and context analysis

One significant flaw in the extension of Heinrich's triangle in Figure 5 is the assumption that by reducing the rate of all the at-risk behaviors, the organization will reduce the chance of incidents higher up the triangle, including FSIs. While behavior is correlated to the occurrence of an incident or injury, *the form and context* of the behavior will ultimately influence the severity of the outcome, not the mere occurrence of behavior. The form and context analysis looks at the topography of responses and the environmental events surrounding the behavior, and enables us to identify possible precursors to future FSIs. For instance, working at a significant height (context) without wearing fall protection (form) is considered an at-risk behavior and an OSHA violation. Using safety glasses (form) when a full face mask is required because of a high amount of fly ash in the work area (context) is also an at-risk behavior and an OSHA violation. They would both be occurrences of at-risk behavior in Figure 5, but the fall protection violation is more likely to relate to an FSI than the eye protection violation. Understanding the form and context of certain behaviors is crucial for accurately assessing risk and preventing serious injuries.

Analyzing the form and context of a work activity is important to understanding and addressing injuries and near misses. For instance, an associate could sustain a significant knee contusion requiring first-aid attention for a variety of reasons. In one instance, the associate was looking at his or her phone on the way to the lunch break (form), and fell down a well-marked, two-step staircase (context). In another, the associate was climbing (form) a poorly maintained railroad tank car (context) and slipped off the curved ladder because lubrication was leaking on the rungs. The form and context of both behaviors in the above incidents are significantly different from each other, but led to the same outcome—injury. They were both first-aid cases, but falling from the oil tanker has a significantly higher likelihood of causing a serious injury (and even a fatality) than falling down two cafeteria stairs. In the first example, the behavior (form) was clearly the more important feature of the accidental fall down a well-marked staircase. In the second example, both form and context played a possible role in the resulting injury.

When looking at all four examples of violations and injuries, Heinrich's triangle (and its derivatives) misrepresents the uniformity of risk between all of the single incidents in each section of the triangle. Because the traditional triangle aggregates each occurrence into the category rows without discriminating the form of the behavior or the context in which it occurred, it cannot be adequately used for the prediction and control of serious incidents. Ultimately the traditional triangle "suggests that the ratios may exist, [but] we cannot predict that we will reduce serious injuries just because we have reduced minor injuries" (McSween, 2015, p. 11). Behavioral scientists can look at form and context as *precursors* to serious incidents; this added dimension will help with prediction and control of safety outcomes.

The traditional triangle can be modified to include the form and context of worker activities as *precursors* of serious incidents. A precursor of serious incidents "is defined as a high-risk situation in which management controls are either absent, ineffective, or not complied with, and which will result in a serious or fatal injury if allowed to continue" (Krause & Murray, 2012, p. 3). According to Wachter and Ferguson (2013), precursors include both "unmitigated high-risk situations" and "high-risk activities." Both can simultaneously occur as "high-risk event combinations" when the form and context of behavior merge to accelerate risk. Focusing on precursors in order to discriminate the differences in these events will assist in the goal of prediction and control of serious incidents.

Adding additional detail to the safety triangle refines the analysis of incidents. In Figure 6, the shaded middle section represents the range of the triangle that includes high-risk situations, activities, or event combinations, all of which may be precursors to an FSI. The dots in the inner triangle represent serious injury precursors, that is, the subset of behaviors, incidents, and hazards in the shaded area could potentially have been serious injuries or fatalities. As such, the events in this subset require special attention from safety professionals, safety committees, and leadership. Notice how the middle section subsumes the two riskier examples from the previous discussion, and the less-risky examples are still on the triangle in the same row, but not within the middle section. This added range makes the incident analysis more robust.

The example of not wearing fall protection at significant height is in the At-Risk Behavior row of the triangle, and also *inside* the middle section of serious incident precursors. This single incident is only in the At-Risk Behavior row because the person did not get hurt, but there was a potential for death or significant injury. Similarly, the example of wearing glasses instead of a full face mask around fly ash is also in the At-Risk Behavior row; however, it is not in the middle section of the triangle, which could be affiliated with potential death. Previously, the traditional triangle treated these two incidents as equivalent. Adding the middle section of the triangle helps discriminate the more precarious at-risk behaviors, and helps highlight the events that need attention for preventing serious incidents.

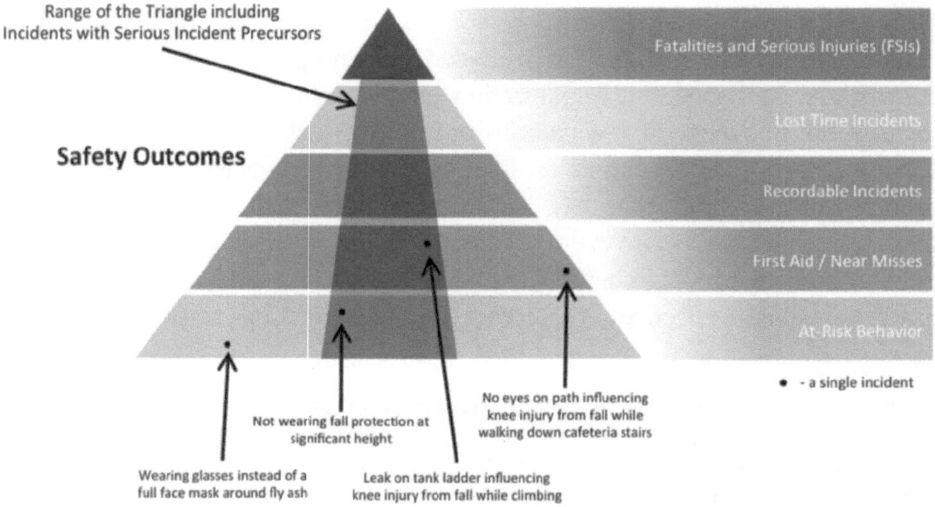

Figure 6. A proposed safety triangle with a third section representing high-risk situations, activities, or event combinations.

The vertical, middle section of the triangle is also pragmatic when analyzing our injury examples. The two dots in the First-Aid/Near Misses row represent incidents requiring attention, but only one is put in the shaded range of the triangle that is related to potential fatality. Of course, professionals would prudently analyze the form and context for both knee-injury incidents; yet, the one that included falling from an oil tank car due to poor maintenance is going to require significantly more attention because it could have led to a significantly worse injury. The traditional triangle implies that falling down two cafeteria stairs and falling off a tank car were equivalent because the outcome was similar (they were both first-aid cases); however, these incidents in the First-Aid row were not equivalent regarding potential for FSIs. Heinrich's triangle is not useful for prediction and control since it views both occurrences as equal in risk for catastrophe. If professionals would like to predict and control the FSIs, this middle vertical section of the triangle needs to be added to the analysis to help discriminate between the incidents in each row that have the potential for death or serious injury. Categorizing and focusing on precursor events, and mitigating their potential to cause FSIs, are crucial aspects for safety management.

Precursors and safety absolutes

Behaviors that are precursors to serious injury are often known to be high-risk activities and are often codified in the company's safety rules (Krause, 2012). Some companies refer to this set of safety rules as their "Safety Absolutes" or other phrases that identify them as the rules that help prevent serious injuries and death. The safety absolutes typically include such practices as fall protection, lockout/tagout, and the use of permits. The safety absolutes should be included on the BBS observation checklists so the BBS process helps identify potential precursors for additional analysis and action planning. To further ensure precursors are addressed, the organization should influence BBS observers to increase the frequency of observations during high-risk tasks. A well-designed BBS process should help employees learn to identify serious incident precursors in their workplace and what they can do to minimize their risk associated with the hazards. It should also help encourage reporting of close calls, especially those related to process safety. These simple steps can alter the typical BBS process to be more effective at preserving life and preventing catastrophic events.

Extending the foundation of the traditional safety triangle

In addition to placing another section in the triangle by assessing the risk of precursors, it is prudent to extend the focus on context by expanding the safety triangle to include other foundational environmental characteristics at the bottom of the triangle. At-Risk Behavior cannot stand alone at the bottom of the triangle. In both a root cause analysis and a functional analysis, a worker's behavior cannot be deemed the sole root cause of an incident. Behavior analysis assumes that behavior is a function of environmental events (Malott & Shane, 2016; Skinner, 1953), and modern root cause analysis posits that incidents have multiple causes—often including unapparent events (Johnson, McSween, & Polluck, 2016). A worker's actions are influenced by so many other variables, so the employee's work context must be analyzed in order to have a true influence on safer actions in the workplace. Because prediction and control of serious incidents is a goal for the safety profession, behavioral scientists need to emphasize the fact that antecedents and consequences significantly influence at-risk behaviors, and these functional stimuli are often governed by the existence of context concerns including operational issues, physical hazards, and other system failures such as poor leadership. When talking to critics, adding form—observable, pinpointed, safe behaviors—to context can make BBS more acceptable and embraceable.

Figure 7 illustrates the distinction between "Process Issues" and "Safety Outcomes." This is important because safety professionals and safety committees need to understand that behavior is part of the safety process. BBS provides a measure of behaviors that are critical to safety outcomes. Observation data is a process measure, not an outcome measure. It is analogous to measures of temperature, flow rates, pressure, and other measures used to ensure that a process remains in control and results in outcomes that are within allowable parameters. Critical behavior within a process must remain within control limits in the same way (Hyten &

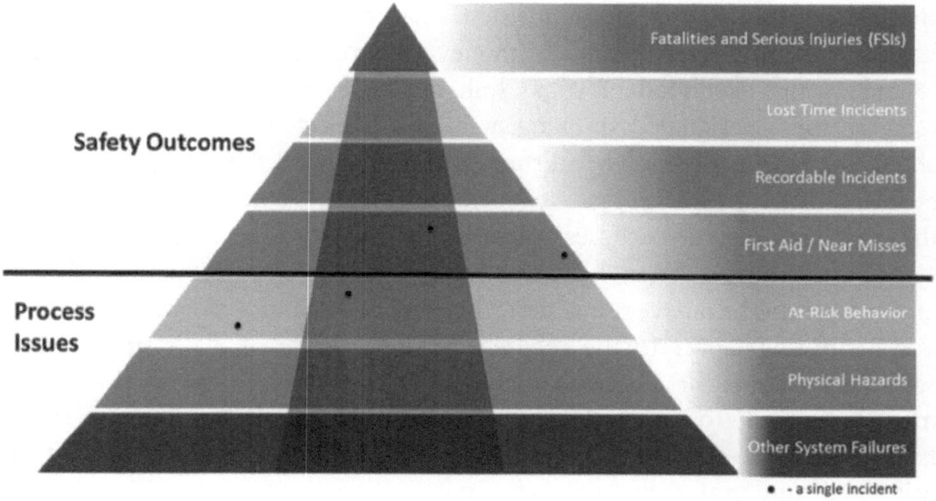

Figure 7. A further extension of the proposed safety triangle that adds systems issues such as physical hazards and other system failures.

Ludwig, 2017). Like temperature and pressure, behavior is a function of the system, but it is not an outcome. Injuries, equipment damage, and barrels of product are outcome measures, but measures of behavior are process measures, not outcomes.

Figure 7 has important implications for addressing and preventing serious injuries by communicating to leaders and safety committees that action plans, especially for serious injury precursors, must go beyond those addressing at-risk behaviors. For incidents in the precursor section of the triangle, action plans should identify and address behaviors by carefully analyzing the other elements of the process represented by different levels of the triangle. Ideally, hazards are eliminated from the work environment, or additional engineering controls prevent the possibility of injury; however, if that is not possible, then other layers of protection should be added.

Physical hazards

An at-risk behavior cannot result in an injury without the presence of a hazardous condition. Leadership and frontline workers have a responsibility to mitigate physical hazards in the workplace. Whether a behavior is at-risk can depend upon context. For example, if we look at the behavior of cigarette smoking, we can say that the action of holding a lit cigarette at the workplace is neutral. It is relatively harmless from a serious incident perspective (cardiovascular health issues for the individual, notwithstanding) when the action is happening in a designated smoking area that is free of flammable debris, visited frequently in a trafficked area, and has commercial ashtray receptacles. The behavior in that context is not at risk for serious incidents because there are no significant physical hazards in the context. At the other end of the spectrum, if the safety inspection system at an oil refinery is faulty, leadership has not spent the right amount of money for procuring the safety equipment to reduce the likelihood of a gas leak during a transfer, ignored workers' requests for a designated smoking area, and neglected to train the workers about the dangers of smoking in some work areas, then the same exact formal behavior of holding a lit cigarette—which was not particularly dangerous in the previous context—becomes an at-risk behavior in this current context.

At-risk behavior often happens because of an environmental context, and if physical hazards are present, then the actions of a worker are simply a contributing factor to the incident. The presence of a physical hazard sets the occasion for a serious incident. The hierarchy of hazard controls must be applied to the context of work to eliminate or minimize risk for FSIs. Often, in certain industrial settings, physical hazards can only be mitigated, and not completely eliminated. Thus, the organization has to move down the hierarchy of controls, making the safety intervention less effective. How leadership, training, procurement, and budget influence both the environmental context and the workers' behaviors is foundational to the entire risk situation (Gravina, et al., 2017; Hyten & Ludwig, 2017; Ludwig, 2017).

Systems and leadership failures

Other system failures establish the base of the expanded safety triangle articulating contributing factors to serious incidents. Many dynamics in industrial settings can set the occasion for physical hazards and at-risk behaviors that could be potential precursors for FSIs. How leaders behave, the effectiveness of training, and the financial resources of a company can contribute to systems failures (Figure 7). When a mechanic fails to learn all of the steps in a lockout procedure, training has failed. This system failure would fall under the shaded area of the "Other System Failures" level of the triangle. When any of these domains are mishandled, they can eventually have a deleterious impact on safety (Ludwig, 2017).

People in leadership positions are still "workers," who engage in behavior having antecedents and consequences, and behavioral science can be used to improve such behavior (Daniels & Daniels, 2007; Houmanfar, Alavosius, Morford, Herbst, & Reimer, 2015; Krapfl & Kruja, 2015; Moran, 2010). Influenced by market forces, production goals, and public relations objectives, leaders often have to make day-to-day choices governed by those contingencies. Oftentimes, those antecedents and consequences influence leadership behavior to be less focused on the primacy of safety goals. When a plant manager is competing with other companies to achieve a record-production year, pursuing the board of directors' expectations to meet aggressive costs containment goals (antecedents), and has a history of financial bonuses for meeting performance goals (consequence), the environment may program the manager up to engage in at-risk behavior. Under such conditions, a manager might cut the safety budgets, reduce overtime in a way than impacts safety training session, and take other steps that may compromise safety or even mechanical integrity. Reducing the effectiveness of training, not procuring proper personal protective equipment (PPE), and mishandling the maintenance of equipment are all system failures. When the plant manager focuses on more proximal and probable reinforcers, those choices contribute to the foundation of this safety triangle. When doing a root cause analysis, professionals attempt to "get to the bottom" of a problem, and a leader's misguided behavior can be a noteworthy system failure functioning as a precursor to FSIs.

How can 21st century BBS assist with FSIs?

As stated previously, form and context function as precursors to FSIs. Identifying precursors requires analysis of form (behaviors) as well as context (process issues). Behavior, including decision making, plays an integral role in process issues as well as safety outcomes. BBS has the potential to make a significant impact on the safety outcomes (e.g., Sulzer-Azaroff & Austin, 2000). Martin and Black (2015) suggest that BBS is a significant and under-used process for addressing serious injuries and fatalities (p. 42). Their observation reflects the fact that most BBS processes do not distinguish between serious injury precursors and other at-risk behaviors. The first problem is that most safety committees managing a BBS process treat all at-risk behaviors with the same analysis and problem-solving methods. The second problem is that action plans to address behavior often do not give adequate consideration to addressing the hazard and other process issues as part of their strategy to address at-risk behaviors. BBS has been shown to be effective for reducing incidents and injuries and "the behavior science community and its industry partners must build on what has been accomplished with personal safety" (Bogard et al., 2015, p. 76). The basic principles and applications of a 21st century BBS can be reorganized to affect change in serious incidents and fatalities.

In brief, BBS has a multistep implementation process (for a more complete description of BBS, see McSween, 2003). Safety professionals work with their safety committee to assess the organization's incidents and injuries, and then operationally define which behaviors need to occur in order to improve safety in that work environment. Those pinpointed behaviors are observed and workers are given feedback about their behavior. As data is collected on safe and at-risk behaviors, the observer provides immediate acknowledgment of safe behaviors, which may prove to be positively reinforcing as they increase in the future. The observer also discusses alternatives for at-risk behaviors with the worker. Those data are aggregated, and after a certain period of time the organization assesses if the behaviors observed meet a certain criteria of safety. If so, the group celebrates meeting that objective. This can serve as a reinforcer as well, and maintains people's safer actions on the job. In addition, trends in at-risk behaviors are analyzed to discover the environmental contexts that set the occasion for dangerous actions. This leads up to management endeavors to create safety action plans to properly address these contextual factors.

Although BBS typically focuses on reducing individual worker injuries, the same assessment-observation-feedback-reinforcement process can be used for reducing FSIs. When doing the assessment, professionals would do well to look at the organization's potential for contextual and behavioral precursors. According to Martin and Black (2015), precursors of serious injuries are identifiable through observations and discussion in 87% of the incidents reviewed. He goes on to describe three types of precursor events: high-risk tasks, high-risk behaviors, and complex or changing circumstances.

The implications for BBS are clear. Those designing a BBS process should study the serious injury precursors and attend to those tasks that could result in FSIs in their organization. The middle section of the triangle can provide an incisive perspective on incidents at each particular workplace. In addition, extending the triangle's base to include physical hazards and other system failures, such as leadership shortcomings, will also assist when looking at the root cause of incidents.

When analyzing the precursors, the BBS process designers should ensure that (a) high-risk tasks are frequently observed, (b) the behaviors critical to those tasks are clearly included on the observation checklist, and (c) the safety committee makes it a priority to review the observation data and identifies precursors—both behavioral and environmental—for additional analysis and intervention. Table 1 presents a comparison of design considerations for serious injuries as compared to BBS efforts to reduce injury frequency. In most BBS processes, the critical behaviors are defined based on data of injuries from the past three to five years. The behaviors are the basis for an observation checklist used for frequent observations in the workplace. When designing a BBS process to prevent serious injuries and fatalities, the critical behaviors often must be identified from a hazard analysis of the workplace, often done by the local safety professional. These behaviors must either be included on conventional BBS checklists, or a separate Serious Incident Prevention (SIP) checklist used for reviewing the specific critical tasks. An additional difference is that the BBS process must focus observation on the hazardous tasks common to fatalities and serious incidents within the facility. For example, in construction, the checklist would often have critical behaviors related to fall protection and scaffolding, and special observations should be taken of those working at heights. In most manufacturing contexts, observations should be routinely performed on jobs requiring lockout/tagout procedures to control energy release.

Furthermore, once the BBS process is in place and providing data from the work samples, the safety committee must communicate all serious injury precursors to the leadership team for review, and an action plan must be recommended. Ultimately, the leadership is responsible for ensuring that serious injury precursors are addressed with an appropriate level of control. If observations document at-risk behaviors related to serious injury

Table 1. A Comparison of Design Considerations for Behavior-Based Safety When Targeting a Reduction in Injury Frequency Versus Prevention of Serious Injuries and Fatalities.

To reduce and prevent	Source of critical behaviors	Observations
Occupational Safety and Health Administration (OSHA) recordable incidents	Identified through analysis of 3–5 years of data	Conducted at random
Serious incidents and fatalities	Hazard analysis by subject matter experts	Conducted during high risk tasks

precursors, leadership must respond with additional layers of protection to ensure that safety practices stay consistent and under control.

BBS safety committees typically analyze observation data to identify where at-risk behaviors are occurring. When these data identify an at-risk behavior that is occurring frequently (number of occurrences) or consistently (percent at-risk) in the previous month's data, they identify a target level of improvement and develop an action plan to increase the related safety practice. The target behavior is not a serious injury precursor according to Figure 8. For this type of target, the action plan might be fairly simple, such as communicating the target behavior, reviewing the topic during safety meetings, and asking observers to increase the frequency of feedback for the target behavior.

Such simple, low-cost interventions are not appropriate for the precursors of serious incidents. If the safety committee identifies an at-risk behavior that is a precursor to a serious incident, both the analysis and the action plan must be more rigorous and complete. Ideally, the analysis of those behaviors and hazards should be addressed through engineering those concerns out of the workplace. When that is not possible, adding additional layers of protection must address the systems issues. For example, in construction, practitioners may not be able to eliminate the task of working at heights, but might ensure a safety professional or supervisor participates in the pre-job briefing and reviews all the controls (such as fall protection, guardrails, and toe boards, etc.) at the beginning of the job, throughout the day, and any time the circumstances change.

As shown in Figure 9, if the safety committee finds *a single instance* of the precursor behavior, they should do a careful functional analysis of the Antecedent, Behavior, and Consequences (an ABC analysis). Then, based on that analysis, develop an action plan addressing the behavior, the hazard, and potential other

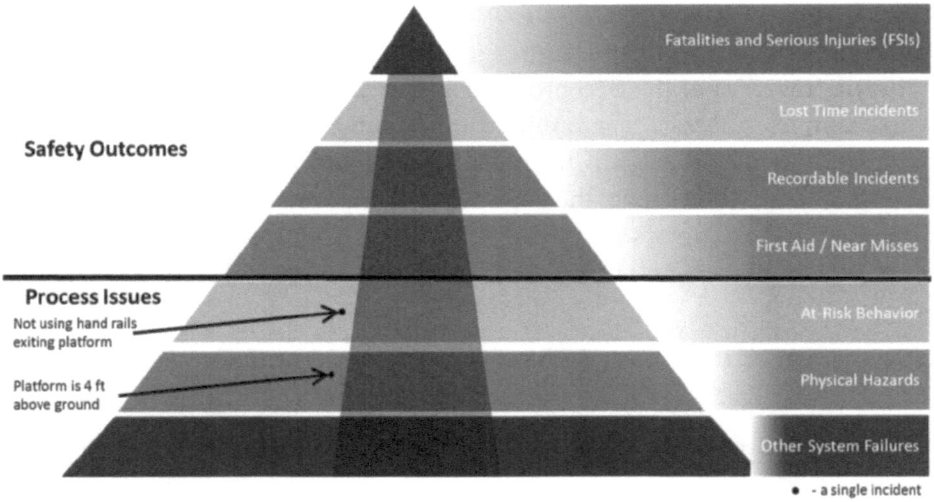

Figure 8. For at-risk behaviors that are not precursors to serious injury, action plans may target improving behavior through feedback, tracking the behavior to ensure that it improves.

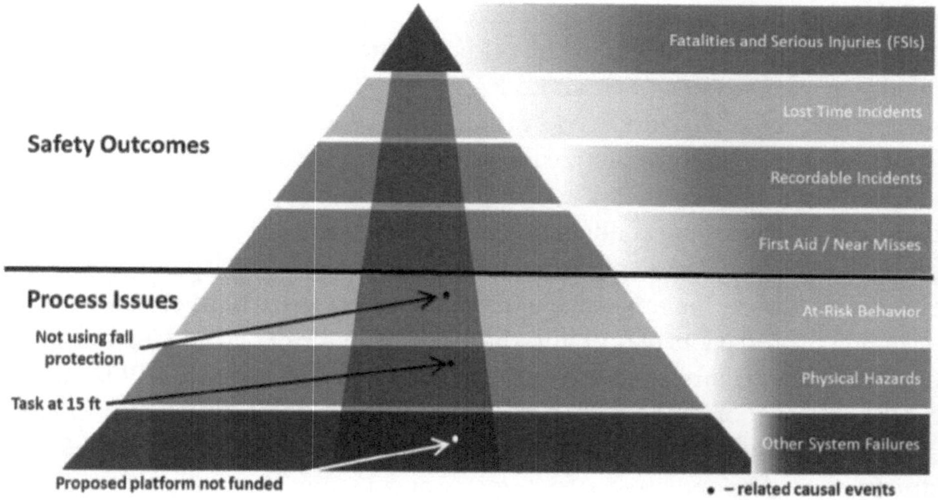

Figure 9. Behavior-based safety can help identify serious injury precursors, such as employees working at heights without fall protection. For precursors, the hazards should be addressed through engineering and additional layers of protection. Action plans for such precursors should address all levels, not just added feedback to address the behavior.

aspects of the process. In the diagram, observations identified that an employee was working on the top of a tank car (at a height or over 15 feet) without fall protection. The action plan should look at ways to eliminate or reduce the hazard. Appropriate interventions might include a permanent or mobile platform, creating better tie-off points for fall protection, or even finding ways to automate the task that do not require climbing on top of the rail car. A more robust analysis might look to see if they have experienced near misses and other historical facts related to the task.

The safety committee's interventions should ensure a level of control appropriate to the hazard or risk in the workplace. Precursors to serious injury often justify the capital expenditures typically associated with robotics or new equipment that eliminate the serious injury hazard. Interventions addressing serious injury precursors and at-risk conditions may require equipment upgrades that reduce the hazards or help support safe behavior. As a final fail-safe, the leadership team should review the analysis, planning, and implementation of actions taken to address serious injury precursors.

Extending these concepts to process safety

Process safety combines managing and engineering skills with an aim to prevent catastrophes. In the petrochemical industry, process safety generally refers to "keeping it in the pipes," which implies ensuring that the facility does not experience any release of product that contaminates the environment or results in a fire or explosion. Maintaining accurate records of process upsets in shift log books, communicating process events at shift change, relaying new leadership decisions, and maintaining accurate process and instrumentation diagrams are all behavioral issues influencing the likelihood of catastrophic events (Rodriguez et al., 2017). All are subject to review through behavioral observations and to the same factors as those shown on the enhanced safety triangle.

Catastrophic process events are admittedly more complex and involve *multiple* causes, often existing at different levels of the organization (Ludwig, 2017). In his book, *Failure to Learn*, regarding the Texas City Refinery explosion, Andrew Hopkins described the unfortunate deviations of critical actions at many levels, including the behaviors of operators, problems with the design and instrumentation, leadership failures, and even regulatory agencies. Hopkins (2010) identifies organizational root causes including "an inappropriately focused remuneration system; cost cutting without regard to safety consequences; an organizational structure that disempowered safety experts; and a senior leadership that discouraged bad news and failed to understand the distinctive nature of process safety" (p. 4). These factors were causal events not just in the Texas City disaster, but also in BP's failure to learn from six close calls in the 10 years prior to the explosion in 2005. Figure 10 presents a simplified way to illustrate some of the factors that might have been identified and addressed from an analysis of the close calls preceding the 2005 explosion, such as an operator overfilling a vessel, explosive vapor release, and budgetary constraints.

Critical behaviors abound, all with their own context and causal factors. Leadership behavior often has a significant contribution to the causal factors. For example, in Hopkins's discussions of the 2005 BP Texas City Refinery disaster, he commented, "A final factor in this story is the failure of leadership at the very

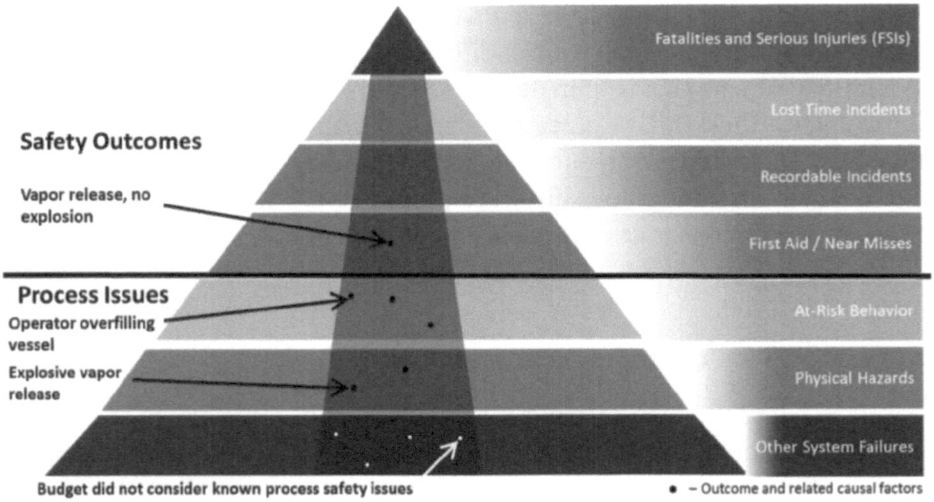

Figure 10. Safety leadership teams should investigate close calls that could have been catastrophic events, identify critical behaviors and hazards, then develop action plans that address engineering and systems issues. Feedback and training alone are inadequate for addressing these issues.

highest level. The CEO was perceived by those around him as unreceptive to bad news about safety. Consequently he was never informed about the deleterious impact of cost cutting at the Texas City site (p. 5)." The same ineffective leadership behavior would retroactively become known as a contributing factor in the 1986 space shuttle Challenger disaster, and again in the 2010 BP Deepwater Horizon oil spill. An emphasis on leadership does not negate the requirements of safe behavior and proper safety engineering, but it simply adds another element to be considered for ensuring serious incident prevention. The challenge for BBS professionals is to apply our technology in ways that address behaviors at all levels, and not just those of the frontline employees.

If the potential for a catastrophic event is identified (whether through behavioral observations, review of near-miss events, or rigorous hazard analysis), organizations should strive to minimize the risk through design. A critical task for behavioral technology is to provide behavioral measures that identify and address problems before catastrophic events occur.

Conclusion

Preventing serious incidents with behavioral science requires proper assessment and intervention. The BBS approach discussed in this paper includes identifying critical behaviors and hazards, incorporating judicious leadership choices, and developing action plans addressing engineering and systems issues. Advanced 21st century BBS implementations are not only broader and more inclusive than the common traditional approaches, but also provide better effectiveness to an organization's safety interventions. When traditional observation and feedback processes are improved so they accelerate change in leadership behaviors and other behaviors critical to process safety, then organizations will have more robust safety successes to celebrate and repertoires to reinforce.

References

American Society of Safety Engineers. (2011). Retrieved January 16, 2016, from http://www.asse.org/asse-says-latest-bls-workplace-fatalities-report-should-be-a-call-to-action

Bogard, K., Ludwig, T. D., Staats, C., & Kretschmer, D. (2015). An industry's call to understand the contingencies involved in process safety: Normalization of deviance. *Journal of Organizational Behavior Management*, *35*, 70–80. Doi:10.1080/01608061.2015.1031429

Bureau of Labor Statistics. (2016, April 21). Revisions to the 2014 Census of Fatal Occupational Injuries (CFOI). Retrieved April 30, 2016, from http://www.bls.gov/iif/cfoi_revised14.htm

Daniels, A. C., & Daniels, J. E. (2007). *Measure of a leader: The legendary leadership formula for producing exceptional performers and outstanding results*. New York, NY: McGraw Hill.

Gravina, N., Cummins, B., & Austin, J. (2017). Leadership's role in process safety: An understanding of behavioral science is needed. *Journal of Organizational Behavior Management*, *37*(3–4), 316–331.

Heinrich, H. W. (1931). *Industrial accident prevention: A scientific approach*. New York, NY: McGraw-Hill.

Hopkins, A. (2010, July). Why BP ignored close calls at Texas City. *Risk and Regulation*, 4–5.

Houmanfar, R., Alavosius, M. P., Morford, Z. H., Herbst, S. A., & Reimer, D. (2015). Functions of organizational leaders in cultural change: Financial and social well-being. *Journal of Organizational Behavior Management*, *35*, 4–27. Doi:10.1080/01608061.2015.1035827

Hyten, C., & Ludwig, T. (2017). Complacency in process safety: A behavior analysis toward prevention strategies. *Journal of Organizational Behavior Management*, *37*(3–4), 240–260.

Johnson, D., McSween, T. E., & Polluck, R. (2016, June 27–29). Applying systems thinking to improve safety. *Proceedings of the ASSE Professional Development Conference and Exposition*, Atlanta, GA.

Krapfl, J. E., & Kruja, B. (2015). Leadership and culture. *Journal of Organizational Behavior Management*, *35*, 28–43. Doi:10.1080/01608061.2015.1031431

Krause, T. R. (2012). New perspectives in fatality and serious-injury prevention. *Presentation at Fatality Prevention Forum 2012*, Coraopolis, PA.

Krause, T. R., & Murray, G. (2012, June 3–6). On the prevention of serious injuries and fatalities. *Proceedings of the ASSE Professional Development Conference and Exposition*, Denver, CO.

Ludwig, T. (2017). Process safety behavioral systems: Behaviors interlock in complex process safety meta-contingencies. *Journal of Organizational Behavior Management*, *37*(3–4).

Malott, R. W., & Shane, J. T. (2016). *Principles of Behavior*. New York, NY: Routledge.

Manuele, F. A. (2002). *Heinrich revisited: Truisms or myths*. Itasca, IL: National Safety Council.

Manuele, F. A. (2013, May). Preventing serious injuries & fatalities: Time for a sociotechnical model for an operational risk management system. *Professional Safety*, 51–59.

Martin, D. K., & Black, A. (2015, September). Preventing serious injuries & fatalities—Study reveals precursors & paradigms. *Professional Safety*, 35–42.

McSween, T. E. (2003). *The values based safety process: Improving your safety culture with behavior-based safety* (2nd ed.). New York, NY: John Wiley.

McSween, T. E. (2015). Preventing serious injuries and how BBS can contribute. *Journal of Applied Radical Behavior Analysis Proceedings of the 9th BBS & OBM European Conference.*

Moran, D. J. (2010). ACT for leadership: Using acceptance and commitment training to develop crisis-resilient change managers. *International Journal of Behavioral Consultation and Therapy*, *6*(4), 341–355. Doi:10.1037/h0100915

RAND Corporation. (2007). *In the name of entrepreneurship? The logic and effects of special regulatory treatment for small business*. S. M. Gates, & K. J. Leuschner (Eds.), Kaufmann-RAND Institute for Entrepreneurship Public Policy. Santa Monica, CA: Author. http://www.rand.org/content/dam/rand/pubs/monographs/2007/RAND_MG663.pdf

Rodriguez, M. A., Bell. J., Brown. M., & Carter. D. (2017). Integrating Behavioral Science with Human Factors to Address Process Safety. *Journal of Organizational Behavior Management*, *37*(3–4), 301–315

Skinner, B. F. (1953). *Science and human behavior*. New York, NY: Macmillan.

Sulzer-Azaroff, B., & Austin, J. (2000, July). Does BBS work? Behavior based safety & injury reduction: A survey of the evidence. *Professional Safety*, 19–24.

United States Department of Labor, Bureau of Labor Statistics (2016) Census of Fatal Occupational Injuries. 2012 Census of Fatal Occupational Injuries (revised data). http://www.bls.gov/iif/oshcfoi1.htm

Wachter, J. K., & Ferguson, L. H. (2013, July). Fatality prevention: Findings from the 2012 forum. *Professional Safety*, 41–49.

Leadership and Crew Resource Management in High-Reliability Organizations: A Competency Framework for Measuring Behaviors

Mark P. Alavosius, Ramona A. Houmanfar, Steven J. Anbro, Kenneth Burleigh, and Christopher Hebein

ABSTRACT

High-reliability organizations (HROs) have emerged across a number of highly technical, and increasingly automated industries (e.g., aviation, medicine, nuclear power, and oil field services). HROs incorporate complex systems with a large number of employees working in dynamic, and potentially dangerous environments. Effectively managing contingencies in HROs, to simultaneously promote safe and efficient behaviors is a daunting task. Crew Resource Management (CRM) has emerged in HROs as a highly effective approach to training and sustaining essential skills within work teams operating across a large workforce. CRM provides a competency framework that enables adherence to standard work instructions while, at the same time, encourages adaptive variance in responding to effectively manage current environmental circumstances that depart from normal routines. This paper considers the development of CRM across several high-reliability industries, develops a behavior analytic account of CRM behaviors, and describes an approach to measuring behaviors within simulated and actual work environments.

The Deepwater Horizon oil rig disaster on April 20, 2010 is a bellwether event in the oil services industry signaling that dire consequences may result when workers in a highly technical, dynamic work setting lose control of complex operations. Eleven people lost their lives that day on the BP managed oil rig in the Gulf of Mexico, and the environmental catastrophe affected millions of people residing along the Gulf coast. The financial impact is estimated to be as high as 60 billion dollars to cover losses and mitigate damage. The Deepwater Horizon Study Group, from the Center for Catastrophic Risk Management (2011), reports that a history of at-risk safety behaviors by workers on the oil rig contributed to the disaster. This event received worldwide attention, which contributed to a renewed focus on human (behavioral) factors that affect operational integrity in increasingly technical, automated contexts where failure can be catastrophic. High-risk industries such as oil field services are becoming increasingly automated but the human elements that remain have the potential to either avert or assist the evolution of a disaster.

High-reliability organizations (HROs) grow from settings in which managerial and operational behaviors are interlocked across levels of the organization in a systematic effort to learn from incidents and accidents and institutionalize corrections to improve safety and performance (Dekker & Woods, 2009). These institutionalized corrections involve highly engineered work settings where processes include redundancies, failsafes, and back-up systems so that nearly error-free technical operations are sustained. In these environments, failures are more likely to result from a complex set of interwoven, organized factors rather than some single component fault. HROs are uniquely challenging work environments as crew members' behaviors interact with automated expert systems to sustain the integrity of processes. HROs include settings that emphasize maintaining operational integrity in the face of challenges inherent in continuous, 24 hour/day operations (e.g., aviation, nuclear power plants, medical settings, and oil rigs). Human factor challenges in these environments may include fatigue due to extended shifts, loss of situational awareness (SA) in the context of complex automated processes with multivariate data streams to guide decisions, and novel contexts that invite behavioral variation from standard workpractices.

HROs utilize humans to monitor data streams, comprehend patterns in complex data, respond to alarms and signals, and sometimes approve or override automated adjustments. Pilots of commercial aircraft monitor preset flight plans run by autopilots and somewhat infrequently control flight maneuvers manually. A similar role is developing for surgeons, train engineers, nuclear power plant operators, oil rig personnel, and other HRO managers where expert technologies conduct many operations. The behavioral challenges in HROs are perhaps unique as the context for work is highly organized and sources of behavioral variance are within the complexity of interlocked contingencies.

Oil service companies aspire to performance levels achieved by the best HROs so that disasters such as the BP event in the Gulf of Mexico are averted (Flin & O'Connor, 2001; IOGP, 2014a, 2014b). A number of high tech industries are leading the way with the development of replicable methodologies to prevent human error and system failure. Transfer of behavioral technologies across these industries offers opportunties to study and understand work behaviors within the context of highly organized processes. Commercial aviation is the exemplar HRO as airlines maintain excellent (in relation to other industries) safety records across millions of hours of operation. Much of this is the result of sophisticated automated flight controls; some is the highly developed competencies of flight crew trained in high-fidelity simulators who learn to step in when automation reaches the limit of its design in a particular situation. The extraordinary example of Captain Chesley Sullenberger, who landed his disabled airliner in the Hudson River with no loss of life, illustrates how humans can fluently interact with technology to solve seemingly insurmountable problems (Sullenberger & Zaslow, 2009).

A challenge to industries aspiring to be top tier HROs is to adopt behavior scientific methodologies into their industry's technologies that resemble those developed in aviation, but systematically adapt the methodologies to the indiocyncracies of their industry. Medicine, for example, is emulating aviation with the developiment of simulators of surgery and patient care (Simon et al., 2000; Zeltser & Nash, 2010). Behavior science provides a coherent framework for identifying sources of behavioral variaton in these settings and helps to develop effective interventions that establish and sustain operational integrity by medical teams. Oil rigs are representative of work in other dynamic contexts such as transportation industries (e.g., commercial aviation, railways), medicine (e.g., surgical teams), and nuclear power (e.g., nuclear power plants) as they integrate highly trained personnel with advanced technologies. Oil rigs are distributed across the planet both on land and at sea and their workforce is often multicultural. Crew members' behaviors required to operate these rigs must interlock as team-level efforts and also engage with off-site personnel (e.g., suppliers, engineers, etc.) who provide ancillary services.

The purpose of this paper is to examine how Crew Resource Management (CRM) has developed as a methodology for managing crew in HROs, and outline a behavior analytic conceptualization of leadership within the context of CRM. We also discuss leadership in the context of interlocking behavioral contingencies, competency in managing behavioral systems, and organizational governance models needed to optimize procedural integrity. CRM offers a useful perspective on leadership and management of HROs as it operates within defined environments (e.g., oil rigs, surgical suites, cockpits, etc.) where the operational boundaries are relatively clear, behaviors can be objectively defined, and the interlocking with other organizational settings (e.g., mission control rooms) can be articulated with a high degree of precision.

Leadership competency for CRM in HRO settings faces two fundamental challenges. First, leaders need to maintain procedural integrity and ensure that crew members adhere to intricate and precise procedures. Control of deviations from desired practice is paramount as teams conduct prescribed duties within defined parameters. On oil rigs, for example, workers follow a prescribed drilling plan that is engineered to reach oil reserves that can be thousands of meters below the Earth's surface. The drilling plan dictates parameters (e.g., depths, pressures) of operations conducted by shifts of workers who drill continuously for weeks or months until the reserve is tapped. Oil rig worker behavior is channeled by rigid parameters that greatly constrain deviation. Second, on occasion, leaders must manage unanticipated events (e.g., excessive pressures) that disrupt well-laid plans. These deviations may escalate to crisis-level events that threaten not just the task at hand, but the lives of the crew and the environment in which they operate. Such was the case on the Deepwater Horizon where the crew lost control of the well and a blowout occurred that could not be contained by engineered fail-safe systems. In these crisis situations, the leader and crew must detect and track the changing context and adapt their responses to address the emerging crisis. Failure to adjust in a timely way can be catastrophic. Thus, CRM entails rigid adherence to standardized work that must also flex to alternative responses when conditions change and standard work is no longer effective. Similar events occur in commercial aviation, nuclear power, and medicine where standard work generally achieves the desired result, but on occasion, conditions emerge that require nonstandard responses to achieve the goal of operations. The behavioral challenges in sustaining routine operations, but also adapting to rapidly changing conditions are considerable (Alavosius, Houmanfar, & Rodriquez, 2005) and call for a competency model that bridges these two fundamental features of CRM (rigid compliance with rules, and flexibility in tracking changing conditions and adapting to them—see Wulfert, Greenway, Farkas, Hayes, & Dougher, 1994 for discussion of rule governed rigidity).

Finally, we present an inventory (Table 1) to illustrate the resources that HROs are applying to enhance crew leader's ability to manage coordinated work. Rules are central to CRM as the crew working in dynamic, high-risk settings may encounter ambiguity due to unexpected events that can be catastrophic if behaviors do not adapt to the changing, complex relationship between stimuli, response choices, and consequences that likely follow each choice. The paper concludes by discussing opportunities for researchers to study these variables affecting leadership and CRM in HROs

Definition of CRM

According to the Industry of Oil and Gas Producers (IOGP, 2014a) "a step-change improvement in operational safety and efficiency of well operations teams (i.e., the full spectrum of drilling, completions, work-overs, and interventions) can be achieved through effective development and application of *nontechnical skills*, also known as Crew Resource Management (CRM)." Crew leaders must not only manage the engineered technical aspects of oil field operations (Flin & O'Connor, 2001) but also interlock the behaviors of team members in light of the competencies and skills of crew members that vary as a function of the work demands. CRM can be viewed as a cascading chain of behavioral events where the leader and crew members effectively utilize available resources to:

(1) plan a work process,
(2) brief everyone on roles/functions,
(3) monitor the process as it occurs,
(4) detect and report deviations from the plan,
(5) communicate corrections from the top down,
(6) adjust actions as needed,

Table 1. Features of CRM and SA in Leading Industries.

CRM/SA effort	Oil field services	Aviation	Medicine	Nuclear
CRM training	Emerging efforts in simulators and CRM training for well site personnel	Advanced: codified and scheduled	Mid-level: beginning to interlock medicine and nursing staff	Advanced: based on aviation and military systems
		Pilot licensing		Train individual and crew competence
		Annual checks		
Simulators	Mid-fidelity: rig floor, drill shack replicas in classrooms	High fidelity: simulated cockpit matches actual aircraft	Mid-high fidelity: surgery suites and treatment rooms	Mid-high fidelity: control room replicas in classrooms with link to external resources
			Mannequins simulate patient	
SA definitions	Cognitive/ contextual	Cognitive framework	Cognitive framework	Cognitive framework
	Developing metrics of competency in context	Foundation work in SA—advanced instrumentation	Individual and group definitions	Individual and group definitions
	BEM identifies sources of variation		Implicit biases being examined	
Behavior metrics in training sites	Emerging competency framework—pre/post tests of knowledge and skills	Multiple measures and rating scales	Cognitive/behavioral measures	Cognitive/ behavioral measures
		Pass/fail score in practical tests		Pre/post tests of knowledge and skills
Behavior metrics in work sites	Unknown	Unknown	Unknown	Unknown
CRM in incident investigations	Unknown	Black box to collect data—open reporting and documentation	Emerging—culture of suppression of reporting	Data records, communications, and logs
Emerging efforts	WSL interlocks with Driller and crew members	Automation, HCI	Implicit attitude and bias in CRM—assess and mitigate	HCI
	Automation, HCI			

Note. CRM = Crew Resource Management; SA = situational awareness; BEM = Behavior Engineering Methodology; HCI = human–computer interaction.

(7) debrief at important moments (at significant change or conclusion of work), and
(8) learn to refine the human-machine interface.

In complex and dynamic situations, CRM orchestrates cooperation among crew members that have different vantage points on the process. Combined, these perspectives optimize adaptive behaviors by all members of the team and may include the input of remote personnel who monitor the process from afar (e.g., from mission control settings). The behaviors involved in CRM can be observed in several key events involved in managing a complex and dynamic process. A chain of CRM behaviors often begins with a briefing meeting during which the team leader informs the crew members of the task ahead, reviews their individual and collective roles and responsibilities, and provides objectives to gage progress. Following the briefing, the crew members conduct the task and work together to assess progress and meet project objectives. Unexpected events may thwart progress and crew members communicate their observations, perhaps in the format of a debriefing meeting, to decide on an adjusted course of action. Upon completion of the task or other significant event (e.g., handoff to another crew at shift change) a debriefing is held to share updates on progress, review actions taken and summarize lessons learned for future operations. These three events (briefing, operations, debriefing) provide useful vantage points for assessors to examine competency by leaders and crew members in context.

CRM is described in the aviation literature as "the effective use of all resources, including hardware, software, and people, to achieve the highest possible level of safety" (Northwest Airlines, 2005). Essentially, CRM can be viewed as a systemic model of training and behavioral change that results in a reduction of human error through the use of all relevant and available resources (Kanki, Helmreich, & Anca, 2010). From its origin in aviation, CRM targets several key processes or skills: situation awareness, communication skills, teamwork, task allocation, and decision making (U.S. FAA, 2004). Among different airlines, this list of key processes or skills differs in organization but all focus on defining optimal interpersonal interactions of crew members working cooperatively within a dynamic environment. To facilitate analysis of critical variations in CRM behavior, these skills can be grouped together into six core skill sets that can be measured as the crew interact with their operational environments:

(1) communication,
(2) situational awareness,
(3) decision making,
(4) teamwork,
(5) management of limits of crew members' capacities, and
(6) leadership.

Each of these six competency domains can be deconstructed into more molecular definitions of behavior within some defined context and viewed within the briefing, operations, and debriefing milestones. Leadership serves an essential integrative function of all elements, without which, CRM would not sustain or lead to desirable outcomes in an organization. Before examining these domains in detail in pursuit of measurable dimensions of these competencies, we briefly review how several HRO industries approach CRM. From this overview, a behavioral account of CRM can be offered.

History and development of CRM

On December 29, 1972, a commercial airliner crashed into the Florida everglades, resulting in 101 fatalities. On March 27, 1977, two commercial airliners collided on a runway at the Tenerife Airport in the Canary Islands, resulting in over 500 fatalities. The Tenerife Airport Disaster holds the record for highest casualties in any single aviation accident. Investigations into these catastrophic events, along with others, revealed a common pattern: an estimated 60–80% of incidents were due to human error. These errors resulted from faulty leadership behaviors from captains (e.g., the breakdown of effective interpersonal communication, and poor decision making) (Wagener & Ison, 2014). Following these findings, CRM was introduced into the aviation industry as a means of addressing these issues and preventing similar incidents.Since its inception in aviation, CRM has been adapted by other high-risk industries. CRM research has generally focused on a particular component, SA, and much of this research is conducted from a cognitive perspective. A review of this research relating to components of CRM is presented below, with a focus on three high-risk industries: aviation, medicine, and nuclear power to illustrate the diffusion of CRM across HROs.

Aviation

The introduction of CRM into aviation facilitated a dynamic culture shift among the crew working in this industry. Historically, the hierarchies within aviation were the driving force behind the organizational culture. A captain was seen as the ultimate authority figure of a flight crew. A captain's word was law, and should never be questioned or challenged. In such hierarchical structures, human error can still occur, as the captain is not exempt from making poor decisions. Therefore, it is crucial that effective, clear, two-way communication exists between the captain and any subordinates to take advantage of all available resources (their knowledge, observations of current conditions, interpretations of data, etc.) that crew members have to offer. CRM's approach to communication has historically been simplified by major airlines using four key words: "*authority* with *participation*, and *assertiveness* with *respect*" (Northwest Airlines, 2005). Throughout airline CRM training, crew members are taught to speak up *before* an incident occurs, rather than to place blame *after* an incident occurs. For instance, if a copilot notices some deviance from standard operating procedures (SOPs), it is their obligation to speak up and it becomes the captain's obligation to take the co-pilot's observation into account.

As previously described, SA is an essential component of CRM. One common feature of training SA in aviation is the utilization of high-fidelity simulators. In some airlines, pilots are required to participate in simulator training as often as every nine months, for a period of two days at a time. At-home computer modules are released on a quarterly basis to complement this hands-on simulator training. During the simulator training, pilots are exposed to a variety of scenarios in which instrument malfunctions, adverse weather conditions exist, or other adverse events are present. Pilots must then respond in ways that incorporate all available information in order to maneuver their simulated aircraft to safety. Their ability to successfully engage in these behaviors are measured and subjectively scored by expert instructors (e.g., using a 1–5 grading scale).

The use of cockpit simulators in the context of SA research is strengthened by the high fidelity of these simulators. Research demonstrates that when using simulator training, the configuration of data needs to be designed in simple ways that make it easily accessible for pilots (Jenkins & Gallimore, 2008), the use of objective measures such as physiological readouts and explicit questions during simulation freezes lead to greater SA (Vidulich, Stratton, Crabtree, & Wilson, 1994), and there may also be some utility in the design and implementation of subjective questionnaires with regard to measuring SA (Waag, 1994).

Systemic application of CRM has made commercial aviation the leading industry in transportation safety, with an average of 0.07 passenger fatalities per billion passenger miles (Savage, 2013). SA training within organizations in high-reliability industries such as aviation allows for the development of a safety culture, whereby the behaviors, practices, policies, and structural components seen within a given organization combine to emphasize increased levels of safety (Meshkati, 1997).

Medicine

CRM was first adopted within the medical field by anesthesiologists and labeled Anesthesia Crisis Resource Management (ACRM; Howard, Gaba, & Fish, 1992). The goal of ACRM was to bring CRM into the medical field and train anesthesiologists to handle crisis scenarios while working on interdisciplinary teams (Howard et al., 1992). ACRM also utilizes high-fidelity simulators for the purpose of training. Simulated operating rooms are large enough to accommodate entire teams (e.g., other physicians, nurses, and technicians) rather than only one person (like a pilot in aviation) and an instructor (Gaba, Howard, Fish, Smith, & Yasser, 2001). While ACRM has a substantial history of use in the medical field, we found no published study to date that empirically demonstrates the impact ACRM has on patient care (Zeltser & Nash, 2010).

MedTeams, an emergency department training program that incorporates CRM procedures into the medical field, is designed to reduce medical errors via the training of different professionals working together in an emergency department (Morey et al., 2002). Similar to CRM, MedTeams identifies system-level variables as a common source of error to be addressed through team-oriented behaviors and communication. In this application of CRM to medicine, resources used to establish competency include classroom instruction, video scenarios, and a 4-hour supervised practicum (Simon et al., 2000). The MedTeams' curriculum shows a strong effect on reducing clinical errors; one study across seven emergency departments showed an average decrease of clinical errors from 30.9% to 4.4% as a result of this training program (Morey et al., 2002).

A more recent application of CRM principles and strategies in the medical field is TeamSTEPPS (Team Strategies and Tools to Enhance Performance and Patient Safety). This protocol was developed by the Agency for Healthcare Research and Quality in collaboration with the Department of Defense. TeamSTEPPS focuses on training four core skill sets to improve overall patient care and safety in the hands of inter-professional teams working in dynamic settings (Epps & Levin, 2015; King et al., 2008):

(1) leadership,
(2) situation monitoring [SA],
(3) mutual support, and
(4) communication.

Research using the TeamSTEPPS methodology demonstrates reductions in medical errors (Cima et al., 2009; Haig & Sutton, 2006; Mann, Marcus, & Sachs, 2006) as well as improved communication and teamwork skills (Turner, 2012; Ward, Zhu, Lampman, & Stewart, 2015). Within the medical field, the majority of the applied SA research tends to focus on the implementation of TeamSTEPPS, as described above. Specific procedures within the TeamSTEPPS training include call outs, check backs, the two-challenge rule, and "CUS" words (Haynes & Strickler, 2014). Call outs involve publicly broadcasting a patient's vital signs, such that all members of a team receive the information at once. Check backs utilize closed-loop communication, whereby the receiver of information repeats key points back to the speaker. The two-challenge rule is where a team member voices a particular concern twice, and if ignored they move up the chain of command with their concern; this promotes the breakdown of hierarchies to the extent that everyone can be heard. "CUS" words, derived from the acronym, "I'm *C*oncerned; I'm *U*ncomfortable; I don't feel like this is *S*afe!" (Haynes & Strickler, 2014) are a way any member of the team can halt an ongoing situation, via communicating concern, so that it can be reassessed. As with CRM, a major focus is on building team communication and breaking down hierarchical barriers.

Maraccini (2016) designed a behavior analytic intervention—to train values and perspective-taking skills—that was combined with TeamSTEPPS-related technology, to create an interprofessional education (IPE) training package for medical and nursing students. They used descriptive analysis methods—from the behavior analytic literature—to compare communication performance prior to and following the completion of the training package during a simulated hand-off task, versus that of a control task. Results demonstrated significant improvements in interprofessional communication accuracy and frequency during patient handoffs, independent of package type.

While much of this research is in its conceptual stage, there are several applied demonstrations of SA training in the medical field. One study that focused on shared, rather than individual, SA involved 500+ hours of observation in surgical settings over a 6-month period with data suggesting that one of the most crucial components to building up SA is the inclusion of explicit, distributed, and timely communication among team members (Gillespie, Gwinner, Fairweather, & Chaboyer, 2013). Behaviors such as self-talk, closed-loop communication, and overhearing conversations were identified as specific ways teams build shared SA.

Nuclear power

As work in nuclear power becomes more automated, increasingly complex technologies guide operators, often in conjunction with parallel monitoring of several data streams. However, even with advanced control systems, safety protocols, and accident mitigation software, human operators are still ultimately responsible for assessment. CRM integration in the nuclear industry has the primary goal of interfacing human, process, and technology elements by optimizing skilled behavior (individual repertoires) within these systems. Hamilton, Kazem, He, and Dumolo (2013) describe a systematic approach to safety and human factors that includes the integration of both engineering and human error data. Within the nuclear industry, human-computer interaction (HCI) and human reliability assessments (HRA) are applied through each component of a project's life cycle. Assessment methods include task analysis to identify critical behaviors, reviews of an operator's experience (behavior repertoires and ability), and functional allocation of resources.

Measuring SA in HROs often includes the use of published measurement tools including direct probe measures like the Situational Awareness Global Assessment Technique (SAGAT) and the Situation Present Assessment Method (SPAM), and qualitative measures like the Situational Awareness Rating Scale (SART) and others (Endsley, 1988, 1995b). These assessments are typically customized to specific scenarios common to the industry and conducted in simulators where raters score crew members' detection of variation in critical features of the work environments. Measurements include latency of responding to changing signals, accuracy of reporting deviations, and comprehension of the consequences of detected variations. Naderpour, Lu, and Zhang (2016) designed and tested the Situational Awareness Support System (SASS) to establish a semiautonomous technology to increase human reliability. The four major elements of this design model include data collection, assessment, recovery of safety failures, and HCI.

Using technology to reduce workload can reduce error and prevent operators from being overwhelmed with data; however, HCI that reduces variability or error in one domain can set the occasion for new types of human error in others (Yang, Yang, Cheng, Jou, & Chiou, 2012). SA and response planning are contextual human factors that reduce error by discriminating critical environmental cues (Naderpour, Lu, & Zhang, 2016). Chase and Bjarnadottir (1992, p. 191) classify two kinds of situations that can produce behavioral variability. First is an environmental change, where behavioral change as a response to that environmental change will produce reinforcement; or second, situations where stimuli changes occur only as a result of behavioral change. With effective discrimination, more precise identification of context and behavior can increase response flexibility. Flexibility training can successfully adapt the human response in ways that control room safety technologies cannot.

Kim and Byun (2011) studied two licensed crew members of nuclear power operators in simulators. Operators were required to mitigate problems by coordinating team members with personnel outside the control room and expert raters scored their individual and team performance using a selection of measurement tools. Their nontechnical competencies were evaluated as attitudes (questionnaires), individual performance (video recording), and collective performance (video recording and simulator records). Results showed that CRM training increased "coordination, communication, and team-spirit" most notable at the level of collective (team) performance. However, the impact was not significant when examined at the level of each individual. This suggests that all (or most) crew members need to be engaged in CRM training, not just the team leader, for positive results to be achieved.

CRM as a behavioral event

Behavior science provides a coherent perspective on the management of human behaviors needed to prevent catastrophe in HROs (e.g., well control on oil rigs). As equipment, automation, and engineering controls become increasingly sophisticated in HROs, the behavior of humans who operate the equipment looms as the most critical factor accounting for catastrophic failure. Behavioral engineering (Gilbert, 1978, 2007) offers a systematic method to identify categories of behavioral variation and institute controls that reduce risk for deviations causing catastrophic losses. The competency of crew leaders and crew members are critical as their behaviors interlock to maintain the integrity of operations, even as conditions change that might undermine the quality and safety of work. The leadership of the crew is a behavioral event that can be examined through the lens of behavior science so that variations are constrained below levels that result in catastrophic failure. Training to fluency aims (Binder, 1996), established by measuring the CRM behaviors of exemplary crew members, offers a structured approach to continuously improving the performance of HROs as managers better understand the behavioral challenges involved in achieving and sustaining effective operations.

Measurement of CRM competencies

Our review of the psychological literature on CRM indicates competency is defined in somewhat different ways across industrial/organizational psychology and behavior analysis. Industrial/organizational psychology might be labeled "the cognitive perspective" and considers competency as a set of enduring traits or characteristics of a person that are emitted across environments. A competent person, from this perspective, is regarded as one who emits adaptive behavior because their cognitive apparatus is sufficiently developed to permit expert behavior regardless of setting. A somewhat different perspective is taken in behavior science ("the operant perspective") where a person is competent if he/she has learned via training and experience to engage in adaptive behavior *within a given context.* Here, competency is assessed within the parameters of a given situation and generalization of skills across settings is not assumed to be automatic. Direct observation of behavior in context defines competency and no effort is made to posit features of the workers' mental facilities.

Both approaches hold merit and convey the notion that competency is a complex set of measures of expert behavior. The cognitive perspective lends itself to selecting workers who fit the task; the operant perspective focuses on assessing supports (job aids, feedback, instructions, training) to enable average workers to become experts. For maximum utility of a competency framework, it needs to consider sources of behavioral deviation if it is to be optimally predictive of adaptive behavior. For example, consider pedestrians walking in New York's Times Square or London's Trafalgar Square. The New Yorker is highly competent to navigate NYC traffic and avoid collision with vehicles; in London, this person might be seriously disoriented by 'foreign" traffic patterns and step in front a lorry. Likewise, the Londoner walks safely through UK traffic but is tentative and uneasy in NYC as swarms of yellow cabs drive on the "opposite" side of the road. The same considerations arise when assessing competency of personnel (e.g., Well Site Leaders [WSLs] on oil rigs). Some basic competencies likely extend across all well sites (e.g., comprehension of drilling parameters) but other critical competencies (e.g., instructing and supervising crew) might be unique to cultural features of geographic locations (a WSL might encounter different forms of crew communications in North Dakota than they would in Chad).

The "soft or nontechnical" behavioral components of CRM are areas of focus across each of the industries reviewed above. Each industry confronts challenges by objectively measuring the technical skills of personnel (e.g., the engineering prowess of field engineers, the "stick and rudder" skills of pilots) and the "human factors" of behavioral interactions of crew members within technical and dynamic environments. Simulators are helpful in measuring the technical and nontechnical competence of individuals and crew members as they react to solve complex system failures. However, more work is needed to examine transference of skills observed in simulators to actual work settings. A competency framework might focus on the individual, but thought should be taken to broaden competency as a collective measure of team performance (collective fluency). Gilbert's (1978, 2007) text *Human Competence* describes Behavior Engineering Methodology (BEM) as a framework for analysis of the sources of behavioral variations seen across these industries and a coherent approach to interventions that increase adaptive behavior by crew members. This methodology identifies approaches to control sources of behavioral variations that threaten operational integrity within the interlocking behavior of multiple workers.

CRM orchestrates cooperation among crew members that have different vantage points on a complex and dynamic process. Combined, these perspectives optimize adaptive behaviors by all members of the team. SOPs are commonly used to guide consistent behavior by crew members and essentially are rules that describe the standard sequence of behaviors and expected results that experts validate as optimal for a given job or task. These rules are antecedent controls and are commonly found to be most effective when consequences (feedback or reinforcement) establish their control of behavior.

Gilbert's (1978, 2007) BEM can be seen as a way to assess crew behavior management in a complex environment. Competency is a critical feature of CRM and in adherence to standardized work where an individual's capabilities must enable them to perform at the level required by the standards. An entry level of proficiency is assumed to be within the repertoire of the oil rig workers (e.g., WSL, Driller, Tool Pusher, Floorman) so that their job aids (tools, instructions, data, etc.) occasion the desired interlocked behaviors under normal operating conditions. A deeper analysis of each of the six domains of CRM leads to assessment of CRM as a chain of behaviors that crew members exhibit as they conduct complex processes, detect and diagnose problems, and adjust operations to maintain procedural integrity. Each of the six competency domains is considered below as building blocks for a comprehensive assessment of CRM competency. For sake of clarity, the six domains are examined arbitrarily as relatively separate response classes. In actuality, the competencies are interwoven as crew members interact with the work context (e.g., the setting, equipment, other personnel, guidance systems, etc.) to complete workflows, detect and diagnose problems, decide on corrective actions, implement solutions, and debrief before the next workflow. Assessments of these competencies by direct observation of behaviors in context (in simulators or at actual worksites) entail assessors measure dimensions of these six competencies during scheduled observation windows. The briefing, operations, and debriefing chronology provides a useful framework in which the observer can view and score behaviors critical in these domains. The competency profile for a given individual would rate observed effectiveness in communication, SA, decision making, teamwork, management of limits of crew members' capacities, and leadership as seen during actual and simulated work events. Competency in the technical domains (e.g., well site engineering) would also be assessed. These six domains are examined below.

Communication

The dynamic nature of HROs necessitates the coordinated behaviors of multiple individuals performing multiple job functions simultaneously to achieve safe and effective results. These results cannot be achieved consistently without effective planning and communication to interlock behaviors. It is important to note that in these dynamic environments, communication is not just top-down, but across levels of a networked workforce. The purpose of targeting communication processes among employees is to break through potential barriers that have been established through hierarchies.

Skinner's analysis of verbal behavior (1957) can be used as a foundation for our understanding of communication. Skinner describes verbal behavior as the behavior of an individual (speaker), which results in the behavior of another individual (listener) interacting with the environment in a manner consistent with the behavior of the first individual (speaker). Reinforcement in such instances is indirect; rather than the direct manipulation of the environment producing reinforcement, the speaker's behavior is reinforced by the listener's direct manipulation of the environment. The speaker and listener roles can fluctuate quite rapidly in everyday life; this is also evident in dynamic, high-risk work environments. With regard to the analysis of communication, a majority of the literature focuses on the topographical characteristics that focus primarily on the verbal behavior of the speaker. In many organizational settings, the dynamic interaction of teams requires our focus on ways by which verbal behavior and its products may affect the listener's behavior. This approach to analysis of language and communication is mainly captured by the functional account of verbal behavior. Relational Frame Theory (Hayes, Barnes-Holmes, & Roche, 2001) provides an empirically supported functional approach to the analysis of language. RFT emphasizes that when we learn to relate events and objects in a certain way, such as comparing them, the function (or meaning/properties) of one event or object transfers to (rubs off on) the other. This transformation of stimulus functions helps to explain why we can say we are happy, for example, when we hear a piece of music that we listened to during a fun winter holiday: the functions of a camp fire, friends, laughter, and food are not only related to the song (i.e., we think about the holiday when we hear the song), but the song has also acquired the enjoyable stimulus function of the holiday (Flaxman, Bond, & Livheim, 2013).

Houmanfar, Rodrigues, and Smith (2009), and Houmanfar and Rodrigues (2012), note that many communications serve to alter the function of stimuli in the workplace, which in turn impacts employee behavior. Rules or statements that change the reinforcing or punishing effectiveness of consequences (in much that same way the establishing operations nonverbally alter the effect of consequences) have been called augmentals (Houmanfar et al., 2009; Maraccini, Houmanfar, & Szarko, 2016; Stewart, Barns-Holmes, Barnes-Holmes, Bond, & Hayes, 2006). Formative augmentals establish a previously neutral stimulus as a reinforcer or punisher. For example, "If we keep costs associated with non-productive time under $100,000 for the month, employees will receive a bonus," will probably result in employees seeking feedback on company expenses, possibly a previously neutral stimulus, and attempting to stay below the specified spending limit. Motivative augmentals, on the other hand, alter the effectiveness of stimuli by altering a consequential function. For example, "Safety is the backbone of our company reputation. If we don't promote safety, we will lose our stature in the industry." This statement takes a stimulus (safety) that already functions as a reinforcer for employees and increases its reinforcing effectiveness.

In high-risk industries, the first instance is the most likely. Some potentially catastrophic change occurs in the environment, and it is up to the crew in that environment to alter their behavior in such a way as to avoid an escalating event. Communication between crew members in such a situation can allow for the generation of motivative augmentals (MOs) that may evoke problem-solving behavior. MOs may be statements that alter the effectiveness of stimuli by altering a consequential function and can be used to increase the importance of team goals while communicating a clear connection between the team actions and goals (Houmanfar et al., 2009)

From a behavior analytic perspective (Houmanfar & Johnson, 2003), team communication can be defined as a psychological event in which team members engage in verbal problem solving under the antecedent and consequential control of an absence of effective rules or presence of heuristic rules. This delineation means that the lack of rules is an antecedent for activity that is itself oriented toward the establishment of such actions (e.g., complete, clear, and accurate specification of organizational contingencies).

As has been discussed conceptually (Houmanfar & Johnson, 2003; Houmanfar et al., 2009) and demonstrated experimentally, no rule and/or implicit rather than explicit (Smith, Houmanfar, & Louis, 2011) and inaccurate rather than accurate (Smith, Houmanfar, & Denny, 2012) rules generate environmental ambiguity. In these studies, environmental ambiguity associated with no rule or ambiguous/inaccurate ones were found to occasion problem-solving behavior among the verbal participants, which in turn led to reduced performance and the self-generation of inaccurate organizational rules on the part of participants. Conversely, implicit (or heuristic) explicit but accurate rules, which minimize environmental ambiguity, were found to produce greater and longer lasting levels of performance.

With regard to HROs, effective communication involves verbal interactions between crew members from all different levels of a hierarchy. Potential systemic barriers to effective communication must be overcome in order to promote safety of a crew. Such barriers include environmental obstacles such as a loud work environment, or interpersonal obstacles such as rigid adherence to hierarchical structures and perceived status granted by such structures for certain roles. In the airline industry, if copilots notice a potential problem, they need to communicate that to the captain, whose responsibility is to then attend to the communicated information and factor that into decision making.

The traditional approach to leadership consisting of chain of command (top-down) approach has to be revisited in the context of CRM. Instead, leaders have to take into consideration the ever-evolving external environment and verbally evaluate the potential adaptations the organization can make to those possible futures. These relations are based on verbally constructed outcomes that, for the leader at least, bears some connection with the current situation. However, these relations must be communicated effectively to the rest of the people in the team if they are to behave in accordance with said relations.

Moreover, individuals' histories of relational networks have a significant influence on the way by which a collectivity of individuals in given interlocked contingencies respond to organizational information generated by each other through communication networks. This interaction between relational networks and communication networks can be captured through the phenomenon of self-organization that is one of the characteristics of social systems (Houmanfar et al., 2009).

Situational awareness

The second component of CRM is SA. SA is described as the ability to monitor or perceive elements in the environment, the comprehension of their meaning, and the projection of their status into the near future. This definition closely approximates the most commonly utilized definition in SA research within clinical psychology: "SA is the perception of the elements in the environment within a volume of time and space, the comprehension of their meaning, and the projection of their status in the near future" (Endsley, 1995a, 1995b). The majority of SA research to date relies on Endsley's foundational concept. While this model comes from cognitive psychology and posits mental processing of information, it can be interpreted in a more explicit manner from a behavioral perspective.

Killingsworth, Miller, and Alavosius (2016) analyze the three components of SA presented in Endsley's model (1995a) in behavioral terms. A brief summary of their analysis is as follows: (a) *perception* can be thought of in terms of stimulus control, conditional discrimination, and observing responses; (b) *comprehension* involves verbal responses relevant to tacting the observable features and underlying functions of different stimuli, as well as their relation to other stimuli and events; and (c) *projection* can be analyzed similarly to predicting, which involves behaving in certain ways based on one's learning history and how that history is interacting with current contingencies (Skinner, 1974).

A number of assessment protocols have been developed to assess SA but the literature offers few published reports on their validation (Durso, Dattel, & Tremblay, 2004; Salmon, Stanton, Walker, & Green, 2006). Despite the relevance of SA to high-risk industries, relatively little objective research has been conducted in regard to measuring the critical behaviors (Craig, 2012). Killingsworth et al. (2016) provide an account of objective behavioral measures of SA that focus on fluency of crew members' behavior in detecting stimulus changes in the work context, communicating these changes to others, and adapting behaviors to meet prevailing conditions. Their analysis and recommendations for behavioral assessments need not be repeated here and readers are referred to that paper for more detail.

Decision making

In complex, dynamic processes anomalies occur that pose choices by operators at critical points to maintain control of the process. For example, an unexpected pattern of data may reveal malfunction or deviation from the prescribed workflow (e.g., unusual pressure readings from the wellbore indicate loss of well control). These set the occasion for follow-up by crew members to investigate and determine the nature of the problem. The diagnosis of the problem entails technical skill in the critical operations (e.g., engineering of the well) as well as human factors in coordination of input from team members leading to collective identification of the problem and mutually agreed upon remediation plan. On oil rigs, the WSL is the equivalent of the captain of a ship and charged with the authority and responsibility to decide on a course of action. How the leader gathers information from the team is important as their unique vantage points on the problem reveal variables affecting decisions. The leader, in light of the team investigations, takes the choice of action with input from crew members that might involve their considerable effort (e.g., leaving their positions, visually inspecting equipment, conducting tests, etc.). Some leaders may act on preconceived definitions of the problem, impulsively decide on a course of action before diagnosis is complete, or otherwise not engage crew members in the decision-making process. In HROs the complex nature of the multiple sources of anomalies warrants collective participation in the assessment and open discussion of the remediation plan.

Risk discounting is a potential threat to decision making (Sigurdsson, Taylor, & Wirth, 2013) especially in complex process malfunctions. Many variables influence decisions, including time pressure, client expectations, probability of negative outcome, delay of possible consequences, (Green & Myerson, 2004) and various interpersonal and culture variables. How problems are framed and described to decision makers alters their discounting curve (Brown & Alavosius, 2014) indicating that language used to convey choices influences decisions. Both the technical competencies of the leader and crew members to ascertain the technical problem and the human factors involved in describing and selecting the optimal solution among a range of options impact the quality of the decision. Making a wise choice is a function of the process by which the team understands the anomaly and their history in regard to risk discounting experiences with similar problems in the past. Training in simulators to solve problems within the stressors likely to be encountered at work sites can establish a repertoire of problem-solving behaviors that adapt to novel situations.

A number of models for decision making have been developed in aviation (U.S. FAA, 1991) and provide a framework for training and assessing competency. DODAR is an acronym for a decision-making process used within aviation that represents the typical process (Moriarity, 2015). DODAR entails a circular flow of steps by which a leader and team members troubleshoot a problem and arrive at a solution. *D*iagnose is the first step to define the problem and what might be causing it. *O*ptions is inventorying the choices and timeframes for corrective action. *D*ecide is to choose the best option available to the crew. This entails discussion and input from crew members. *A*ct or assign is the leader designating appropriate actions to be taken by crew members. *R*eviewing is monitoring the plan and expected outcomes. If the plan is not resolving the problem, DODAR is started again until the solution is reached. An assessor can observe a crew encountering a problem within a simulated workflow and score the speed and accuracy of individual and collective behaviors as the team completes DODAR through resolution of the anomaly.

Teamwork

Teamwork is the coordinated individual and collective behaviors of a crew. One focus of CRM is the breakdown of hierarchical barriers among crew members. This break down allows for more effective coordination of cooperative behaviors as it levels the playing field, so to speak. A core tenet of CRM is that crew members work together toward a common goal (i.e., teamwork). When analyzing the coordinated behaviors of a crew, it is important to distinguish between instances of crew members behaving with respect to outcomes that benefit only themselves versus outcomes that benefit the entire team.

A unit of analysis for teamwork can be found in the metacontingency (Glenn, 1988, 2004; Houmanfar, Rodrigues, & Ward, 2010; Malott & Glenn, 2006; Smith, Houmanfar, & Louis, 2011). Metacontingencies describes selective contingencies that operate on interlocked patterns of behavior between one or more persons or groups of persons. When the behavior of one person (e.g., a crew member at a well site) becomes interlocked with (i.e., dependent upon) the behavior of another, a pattern of behavior emerges, which Glenn (2004) is described as interlocking behavioral contingencies (IBCs). IBCs, when they occur, have a measureable effect on the aggregate outcome (e.g., drilled well). In other words, "metacontingency holds that interlocked behaviors of members constituting a group are selected by the shared environmental consequences they produce for the group members." (Smith et al., 2011). The complexity of the interactions necessitate definitions of the individual and collective behavior classes in context (Bar-Yam, 1997) that combine to form teamwork. The analyses require scaling measures from parts (individuals) to wholes (teams) so that the actions are understood as a coordinated, interlocked behavioral system operating in some organized context.

The role of language is crucial in a metacontingency (Blakely & Schlinger, 1987) but insufficient to understand teamwork, as individual and collective behaviors are coordinated by rules, but members must also track consequences resulting from actions. As teams adapt to changing conditions, the rules that govern their interactions and their learning histories in similar environments are better or worse at enabling adaptive behavior. It is the interlocking contingencies of individual behaviors that account for the metacontingency and its influence on the team outcome.

Management of limits of crew members' capacities

Increasingly complex technology guide operators in HROs often in conjunction with parallel monitoring of multiple data streams; however, even with advanced control systems, safety protocols, and accident mitigation software, human operators are still ultimately responsible for assessment. Unexpected events can be catastrophic in HROs. Prevention focuses investigators to examine condition chains, including interlocked behavior within work flows, how humans fail to sustain vigilance for extended periods, and other threats to operational integrity. Using technology to reduce workload can reduce error and prevent operators from being overwhelmed with data; however, computer-based HCI that reduce variability or error in one domain can set the occasion for new types of human error in others. (Yang et al., 2012). Stress, fatigue, habituation and other factors limit humans' ability to sustain optimal performance. When workflow is paced by automated controls, the demands on humans may impede how they self-manage their performance. The human response to overwhelming demands is unpredictable with reactions like withdrawal, escape, inflexible repetition of ineffective behavior, and avoidance of further stimulation being dysfunctional responses.

Training an adaptive, flexible response can be difficult when operators adhere to standard work instructions but then must detect and respond to novel environmental variations. This is compounded when responses of team members are interlocked to meet demands. Human errors may occur as near misses that increase risk but do not result in negative consequences. Errors, in these cases, may remain below detection and notification threshold and are often unreported. These events are resource intensive to investigate and without detectable consequence (Preischl & Hellmich, 2016) so opportunities to learn prevention is lost. There is value in examination of near misses as they can reveal undetected sources of variation. When design itself is prone to unknown error, operators may repeatedly encounter unpredicted stimuli without contacting a consequence. They may habituate to those stimuli (McSweeney, 2004;

McSweeney & Swindell, 2002) and not respond effectively to important signals. This has been termed "normalization of deviance" (Vaughn, 1996) and is a common feature of suboptimal work behavior learned by workers as they acclimate to the work setting. Habituation can also occur to established alarms as when a worker initially reacts to a warning signal in the environment, but over time the signal no longer occasions a response. A worker may fail to respond to an H^2S alarm despite being trained in the danger and protective action during drills, as there is no consequence for not responding if the alarm is set to blare at low concentrations. If the signal is not paired with a consequence to maintain the discriminative properties that control responses of survival, then stimulus control is lost. This helps explain why there are workers who complete training and understand the risks, but then deviate in the behavior in the field.

Flexibility training can successfully adapt the human response in ways that control room safety technologies cannot. Kim and Byun (2011) studied two licensed crew members of nuclear power operators in simulators. Operators were required to mitigate problems by coordinating team members with personnel outside the control room and expert raters scored their individual and team performance using a selection of measurement tools. Their nontechnical competencies were evaluated as attitudes (questionnaires), individual performance (video recording), and collective performance (video recording and simulator records). Results showed that CRM training increased "coordination, communication and team-spirit" most notable at the level of collective (team) performance. This is interesting as the impact was not significant when examined at the level of each individual. It suggests that all (or most) crew members are to be engaged in CRM training, not just the team leader, for results to be seen.

Carvalho, Benchekroun, and Gomes (2012) identify proactive process as key to the resilience of a nuclear power plant disrupted by variations. SAis maintained and validated through information dissemination occurring during shift changeovers. This is parallel to patient handoffs in the medical field where information exchange during shift changeover reduces error. Danielson, Alvinius, and Larsson (2014) report that SA establishes a common operating picture using values, routines, and rules to set context and prioritize communication. They found that distant proximity and communication barriers can create a disjointed SA while sufficient, timely dissemination of information can expand it. Communicating important information about dynamic environments via briefings and debriefings is critical to effective SA by teams working in complex systems.

Assessment of verbal communications entails measuring the speaker and listeners' behavior. These verbal exchanges can be coded along dimensions of specificity/ambiguity, completeness, accuracy, clarity and other indices of information transfer. Data on communications are taken through logbooks, formal and informal verbal exchanges, information panel walkthroughs, and documented transformation adjustments to scheduled plans. Some of these data streams are collected by the control systems; some are collected by direct observation (live or video) of social interactions. It is instructive to note that competency in a collective situation can be measured at the individual and team level and Kim and Byun (2011) reported the most significant results were seen at the team level where the context of CRM is the team members within a control room.

Crew leadership

As mentioned earlier, CRM necessitates the support of leadership to be effective in that hierarchical leaders are in the position to organize and manage the resources available for a job site. By identifying some key outcome measure (e.g., rates of injury), the safety performance of a leader can be assessed by the performance of a crew. Effective leaders result in effective crew members who produce desired results. Using crew safety as an example, effective crew leaders will design and manage contingencies in a work setting such that there are few, if any, injuries in that particular environment. An effective leader, therefore, facilitates the expected performance of their team (Abernathy, 1996, 2000, 2009; Houmanfar et al., 2009).

According to Houmanfar and Rodrigues (2012), the performance of teams relies on several factors that managers can alter in order to facilitate team success. Performance tends to be better when team members share the same goals and are committed to the same task. Establishing shared goals among team members can be accomplished by laying out a clear vision for the team that is in alignment with the goals and mission of the organization. Specifying clear rewards or outcomes that will arise from team success can also aid in increasing the reinforcer value of the team goals. The team will be more committed to the task if the goals are deemed important and team members believe that they can be achieved. If necessary, augmentals can be used to increase the importance of team goals while laying out a clear connection between the team actions and goals—placing feasible team behaviors in frames of before-after with the team goals—may help increase belief in the possibility of accomplishing the team goals.

It is important to note that leadership in this context is not restricted to organizational leaders; while these individuals are important to the systemic implementation of CRM, any member of a crew can demonstrate leadership. Krapfl and Kruja (2015) identify two key features of leadership: the term itself describes a wide variety of behaviors, and the context in which leadership is observed must be accounted for. A detailed account is given of a "Leadership Behavior Menu," in which behavioral factors of leadership are identified that can be applied in a wide variety of contexts by many different individuals. Some of these leadership behaviors overlap with the most important skills of CRM; therefore, we can see this menu as a list of competencies for leaders. Components of this menu are mapped onto the critical skills of CRM below.

Crew leaders' foci

Collective SA

While not explicitly discussed by Krapfl and Kruja (2015) as a component of leadership, SA can be related to their description of execution skills. Essentially, this describes the ability of crew leaders to make things happen. This does not solely refer to the planning stages of some project or task. Rather, this involves seeing the project or task through to completion by engaging the team throughout the process. Leaders in this regard do not sit back and let others execute the plan for them. Instead, they are involved in each stage of the plan's implementation. It is during this critical stage that SA is essential. The execution of plans will inevitably deviate off course. Effective leaders will perceive, comprehend, and project the changes in a given situation and help guide the team back on course. They create and convey a common picture of the situation so all crew members are aligned. At times, external forces (e.g., demands from clients) may question the integrity of operations and threaten to disrupt teamwork. The effective leader responds to such intrusions and defends the crew members from distractions so that the crew manages the integrity of the process underway.

Leadership of teamwork

Team-building skills and enabling skills (Abernathy, 1996; Houmanfar, Alavosius, Morford, Herbst, & Reimer, 2015; Krapfl & Kruja, 2015) are leadership behaviors that focus on teamwork. Team-building skills involve the selection of effective team members and are therefore necessarily directed toward upper-level leaders in an organizational hierarchy. In CRM the leader is responsible for ensuring that all members of the crew have the competencies necessary to perform their particular task. To do this, they need to utilize the resources that characterize the roles that perform the task and measure the capacity of the team members performing a particular task to ensure that each crewmember is qualified for the role that they fill. This can be explicitly verified during briefing events where the leader confirms that all crew members are present, ready, and able to complete the upcoming tasks. Enabling skills, on the other hand, involve fostering opportunities for team member growth. Many different members of a team can accomplish this. Leaders, in that regard, can come from any level of the organization. Delegation, support, guidance, and feedback are all achievable by each member on a collaborative team. When members of a team work in coordination with a common goal, they create interlocking contingencies that are reinforced at the group level. This interlocked behavior, by definition, puts members of a crew into contact with other members of a crew, which then may create collective contingencies for the tasks being performed.

CRM, leadership, social validity, and HRO culture

Krapfl and Kruja (2015) conclude that the establishment of an organizational culture is "more influenced by the leader than any other single factor." In order to build and sustain a culture of safety, for instance, leaders within HROs must demonstrate the importance of this focus through their own behavior. A leader's emphasis on the importance of CRM, for instance, helps create a safety culture in which organizational members are more likely to adhere to safer behaviors both individually and in team contexts. This is in part due to the nature of CRM as a system of targeting SA, communication, and teamwork at the individual and group level.

Team leaders within HROs have the opportunity to create a work environment in which prosocial behavior is highly valued. Pro-social behavior in these industries can range from ensuring the safety of members of a crew to minimizing or avoiding any detrimental outcomes of work processes on the environment (i.e. a catastrophic oil spill or fatal airplane crash). Recall reports of US Airways flight 1549 being struck by a flock of birds, resulting in the malfunction of both engines. Captain Chesley "Sully" Sullenberger's exemplary leadership resulted in a controlled water landing that saved the lives of all 155 on board the plane. During the ensuing investigations, it was determined through flight simulations that any other alternative course of action would have resulted in mass casualties and destruction of property. The individual and collective behavior of the captain and crew members define exemplary responding of the flight crew in a crisis situation and set performance aims, which might be established in training other flight crew.

Houmanfar et al. (2015) identify three key features of prosocial behavior: "(a) operating in the context of positive reinforcement contingencies for others, (b) minimizing aversive or coercive conditions and contingencies of others while not explicitly operating as *part* of those conditions or contingencies, and (c) aiding others in identifying or achieving optimal levels of choice." On an oil rig, the WSLs, for instance, are in the position to create contingencies supportive of safety behaviors for their team that meet this definition by (a) reinforcing adherence to SOPs during typical working conditions or to acceptable behavior variation leading to beneficial outcomes during a potential crisis event; (b) providing feedback for erred performance in a nonaversive manner, even allowing team members the opportunity to correct mistakes rather than punishing behavior; and (c) providing opportunities for behavioral variation when plausible.

Conclusion

CRM is a systemic intervention that focuses on the reduction of human error through training and behavioral change, while utilizing contextual resources available to assist in these aims. Many different organizations across high-reliability industries have proposed their own account of what skills constitute effective

CRM. We posit that these skills can be grouped together into six broad behavioral classes: communication, SA, decision making, teamwork, managing human capacity, and leadership. Each of these behavioral classes were examined from a behavioral perspective, and relevant analogous research was discussed.

Review of the literature indicates that many HROs are developing expertise in behavior science applications to enhance CRM. The initiatives reveal common directions and areas for cross industry dissemination. Table 1 illustrates the status of key features of CRM in the selected industries reviewed in this paper.

Aviation is the exemplar industry with the most mature CRM processes enabled by high-fidelity simulators and a well-established competency framework to assess crew capability. The nuclear power industry also has an impressive safety record with advanced training and simulators. Medicine is challenged by a culture of under-reporting, likely as a result of malpractice issues, with uncertain metrics on patient safety and medical error. Medicine is pursuing CRM and looks to aviation and military applications for inspiration. Oil field services are initiating formal CRM training and developing a competency framework to gauge the ability of WSLs for CRM, although, rig simulators are relatively low fidelity when compared to those used in aviation.

Effective training and maintenance of these CRM and leadership behavioral classes will help create and maintain an organizational culture with an emphasis on respect and safety. Additionally, the effective implementation of CRM allows organizations to address prosocial behavior in a more effective manner; from keeping their own employees safe to preventing major catastrophes with wide-scale environmental impact.

The role of leadership in creating and fostering an organizational culture, which effectively utilizes CRM, is essential. From organizational leaders' power to implement change across an entire system down to instances of leadership from a member of a crew in promotion of CRM, the function of leadership is essential to sustaining the shift to (and continued use of) CRM in any given organization. Such a shift is especially critical in high-risk industries, as promising results have been demonstrated by those leading industries implementing CRM: aviation, medicine, and nuclear power. The oil field services industry is poised to follow suit and create work environments that are safer for both their employees and the environment in which they operate. As with other high-risk industries, oil field services span the globe, therefore, the potential impact is quite substantial.

The focus on crew behavior in the context of CRM is a useful approach for behavior analysis. By analyzing communication networks, instructions (e.g., SOPs) and other rules, verbal coordination of behaviors, accuracy and latency of responding, and the interlocked behaviors seen in team dynamics, we can objectively measure skills of CRM. By adding objective measurement elements to CRM skills, behavior analysis is extending the conceptual literature from cognitive psychology. Analyzing key behaviors in their context allows us to develop measures of competency that can be used to inform organizational leaders with respect to training, employee feedback, policy changes, and other pragmatic adjustments to the work environments in HROs.

References

Abernathy, W. B. (1996). *Sin of wages*. Memphis, TN: PerfSys Press.

Abernathy, W. B. (2000). *Managing without supervising: Creating an organization-wide performance system*. Memphis, TN: PerfSys Press.

Abernathy, W. B. (2009). Walden two revisited: Optimizing behavioral systems. *Journal of Organizational Behavior Management, 29*, 175–192. doi:10.1080/01608060902874567

Alavosius, M. P., Houmanfar, R., & Rodriquez, N. J. (2005, November). Unity of Purpose/ Unity of Effort: Private-sector preparedness in times of terror. *Disaster Prevention and Management, 14*(5), 666–680. doi:10.1108/09653560510634098

Bar-Yam, Y. (1997). *Dynamics of complex systems*. Reading, MA: Addison-Wesley.

Binder, C. (1996). Behavioral fluency: Evolution of a new paradigm. *The Behavior Analyst, 19*, 163–197.

Blakely, E., & Schlinger, H. (1987). Rules: Function-altering contingency-specifying stimuli. *The Behavior Analyst, 10*(2), 183–187.

Brown, T. W., & Alavosius, M. P. (2014). Language and discounting behavior (unpublished doctoral dissertation - Alavosius Advisor). University of Nevada Reno, Reno, NV.

Carvalho, P. V. R. D., Benchekroun, T. H., & Gomes, J. O. (2012). Analysis of information exchange activities to actualize and validate situation awareness during shift changeovers in nuclear power plants. *Human Factors and Ergonomics in Manufacturing & Service Industries, 22*(2), 130–144. doi:10.1002/hfm.20201

Chase, P. N., & Bjarnadottir, G. S. (1992). Instructing variability: Some features of a problem-solving repertoire. In S. C. Hayes, & L. J. Hayes (Eds.), *Understanding verbal relations* (pp. 181–193). Reno, NV: Context Press.

Cima, R. R., Kollengode, A., Storsveen, A. S., Weisbrod, C. A., Deschamps, C., Koch, M. B., & Pool, S. R. (2009). A multidisciplinary team approach to retained foreign objects. *Joint Commission Journal on Quality Patient Safety, 35*(3), 123–132. doi:10.1016/S1553-7250(09)35016-3

Craig, C. (2012). Improving flight condition situational awareness through human centered design. *Work, 41*, 4523–4531. doi:10.3233/WOR-2012-0031-4523

Danielsson, E., Alvinius, A., & Larsson, G. (2014). From common operating picture to situational awareness. *International Journal of Emergency Management, 10*(1), 28–47. doi:10.1504/IJEM.2014.061659

Deepwater Horizon Study Group. (2011). *Investigation of the macondo well blowout disaster*. Retrieved from http://ccrm.berkeley.edu/pdfs_papers/bea_pdfs/dhsgfinalreport-march2011-tag.pdf

Dekker, S. W., & Woods, D. W. (2009). The high reliability organization perspective. In E. Salas, & D. Maurino (Eds.), *Human factors in aviation* (2nd ed., pp. 123–146). New York, NY: Wiley.

Durso, F. T., Dattel, A. R., & Tremblay, S. (2004). SPAM: The real-time assessment of SA. *A Cognitive Approach to Situation Awareness: Theory and Application, 1*, 137–154.

Endsley, M. R. (1988, May). *Situation awareness global assessment technique (SAGAT)*. Aerospace and Electronics Conference, 1988. NAECON 1988. Proceedings of the IEEE 1988 National (pp. 789–795). Dayton, OH: IEEE.

Endsley, M. R. (1995a). Toward a theory of situation awareness in dynamic systems. *Human Factors, 37*, 32–64. doi:10.1518/001872095779049543

Endsley, M. R. (1995b). Measurement of situation awareness in dynamic systems. *Human Factors, 37*, 65–84. doi:10.1518/001872095779049499

Epps, H. R., & Levin, P. E. (2015). The TeamSTEPPS approach to safety and quality. *Journal of Pediatric Orthopedics, 35*(5), S30–S33. doi:10.1097/BPO.0000000000000541

Flaxman, P. E., Bond, F. W., & Livheim, F. (2013). *The mindful and effective employee: An Acceptance & Commitment Therapy training manual for improving well-being and performance*. Oakland, CA: New Harbinger Publication, Inc.

Flin, R., & O'Connor, P. (2001). Applying crew resource management in offshore oil platforms. In E. Salas, C. A. Bowers, & E. Edens (Eds.), *Improving teamwork in organizations: Applications of resource management training* (pp. 217–233). Hillsdale, NJ: Erlbaum.

Gaba, D., Howard, S., Fish, K., Smith, B., & Yasser, S. (2001). Simulation-based training in anesthesia crisis resource management (ACRM): A decade of experience. *Simulation & Gaming, 32*(2), 175–193. doi:10.1177/104687810103200206

Gilbert, T. F. (1978, 2007). *Human competence: Engineering worthy performance*. Publication of the International Society for Performance Improvement. San Francisco, CA: Pfeiffer.

Gillespie, B. M., Gwinner, K., Fairweather, N., & Chaboyer, W. (2013). Building shared situational awareness in surgery through distributed dialog. *Journal of Multidisciplinary Healthcare, 6*, 109–118. doi:10.2147/JMDH.S40710

Glenn, S. S. (1988). Contingencies and metacontingencies: Toward a synthesis of behavior analysis and cultural materialism. *The Behavior Analyst, 11*, 161–179.

Glenn, S. S. (2004). Individual behavior, culture, and social change. *The Behavior Analyst, 27*, 133–151.

Green, L., & Myerson, J. (2004). A discounting framework for choice with delayed and probabilistic rewards. *Psychological Bulletin, 130*, 169–792. doi:10.1037/0033-2909.130.5.769

Haig, K., & Sutton, S. (2006). SBAR: A shared mental model for improving communication between clinicians. *Joint Commission Journal on Quality Patient Safety, 32*(167–175). doi:10.1016/S1553-7250(06)32022-3

Hamilton, W. I., Kazem, M. L. N., He, X., & Dumolo, D. (2013). Practical human factors integration in the nuclear industry. *Cognition, Technology & Work, 15*(1), 5–12. doi:10.1007/s10111-012-0213-z

Hayes, S. C., Barnes-Holmes, D., & Roche, B. (2001). *Relational frame theory: A post-Skinnerian account of human language and cognition*. New York, NY: Guilford Press.

Haynes, J., & Strickler, J. (2014). TeamSTEPPS makes strides for better communication. *Nursing, 44*(1), 62–63. doi:10.1097/01.NURSE.0000438725.66087.89

Houmanfar, R., & Johnson, R. (2003). Organizational implications of gossip and rumor. *Journal of Organizational Behavior Management, 23*, 117–138. doi:10.1300/J075v23n02_07

Houmanfar, R. A., Alavosius, M. P., Morford, Z. H., Herbst, S. A., & Reimer, D. (2015). Functions of organizational leaders in cultural change: Financial and social well-being. *Journal of Organizational Behavior Management, 35*, 4–27. doi:10.1080/01608061.2015.1035827

Houmanfar, R. A., & Rodrigues, N. J. (2012). The role of leadership and communication in organizational change. *Journal of Applied Radical Behavior Analysis, N1*, 22–27.

Houmanfar, R. A., Rodrigues, N. J., & Smith, G. S. (2009). Role of communication networks in behavioral systems analysis. *Journal of Organizational Behavior Management, 29*, 257–275. doi:10.1080/01608060903092102

Houmanfar, R. A., Rodrigues, N. J., & Ward, T. A. (2010). Emergence and metacontingency: Points of contact and departure. *Behavior and Social Issues, 19*, 78–103. doi:10.5210/bsi.v19i0.3065

Howard, S., Gaba, D., & Fish, K. (1992). Anesthesia crisis resource management training: Teaching anesthesiologists to handle critical incidents. *Aviation, Space, and Environmental Medicine, 63*(9), 763–770.

IOGP. Report 501. (2014a). *Crew resource management for well operations teams.* Project commissioned by OGP's Safety Committee and the Well Experts Committee to the University of Aberdeen. International Association of Oil & Gas Producers.

IOGP. Report 502. (2014b). *Guidelines for implementing well operations crew resource management training.* Well Experts Committee. Training, Competence & Human Factors Task Force. International Association of Oil & Gas Producers.

Jenkins, J. C., & Gallimore, J. J. (2008). Configural features of helmet-mounted displays to enhance pilot situational awareness. *Aviation, Space, and Environmental Medicine, 79*(4), 397–407. doi:10.3357/ASEM.2195.2008

Kanki, B., Helmreich, R., & Anca, J., & ScienceDirect (Online service). (2010). *Crew resource management* (2nd ed.). Amsterdam, Boston: Academic Press/Elsevier.

Killingsworth, K., Miller, S. A., & Alavosius, M. P. (2016). A Behavioral interpretation of situational awareness: Prospects for organizational behavior management. *Journal of Organizational Behavior Management, 36*(4), 301–321. doi:10.1080/01608061.2016.1236056

Kim, S. K., & Byun, S. N. (2011). Effects of Crew Resource Management training on the team performance of operators in an advanced nuclear power plant. *Journal of Nuclear Science and Technology, 48*(9), 1256–1264. doi:10.1080/18811248.2011.9711814

King, H. B., Battles, J., Baker, D. P., Alonso, A., Salas, E., Webster, J., ... Salisbury, M. (2008). TeamSTEPPS: Team strategies and tools to enhance performance and patient safety. In K. Henricksen, J. B. Battles, M. A. Keyes, & M. L. Grady (Eds.), *Advances in patient safety: New directions and alternative approaches (Volume 3: Performance and Tools).* Rockville, MD: Agency for Healthcare Research and Quality.

Krapfl, J. E., & Kruja, B. (2015). Leadership and culture. *Journal of Organizational Behavior Management, 35*, 28–43. doi:10.1080/01608061.2015.1031431

Malott, M. M., & Glenn, S. S. (2006). Targets of intervention in cultural and behavioral change. *Behavior and Social Issues, 15*, 31–56. doi:10.5210/bsi.v15i1.344

Mann, S., Marcus, R., & Sachs, B. (2006). Lessons from the cockpit: How team training can reduce errors in L&D. *Contemporary OB/GYN, 51*, 34.

Maraccini, A. M. (2016). *Examining the impact of an interprofessional education training package on communication during handoff performance in medical and nursing students: A Behavior analytic approach to assessment and intervention* (unpublished doctoral dissertation--Houmanfar Advisor). University of Nevada Reno, Reno, NV.

Maraccini, A. M., Houmanfar, R. A., & Szarko, A. (2016). Motivation and complex verbal phenomena: Implications for organizational research and practice. *Journal of Organizational Behavior Management, 36*, 282–300. doi:10.1080/01608061.2016.1211062

McSweeney, F. K. (2004). Dynamic changes in reinforcer effectiveness: Satiation and habituation have different implications for theory and practice. *The Behavior Analyst, 27*(2), 171–188.

McSweeney, F. K., & Swindell, S. (2002). Common processes may contribute to extinction and habituation. *The Journal of General Psychology, 129*, 364–400. doi:10.1080/00221300209602103

Meshkati, N. (1997, April). *Human performance, organizational factors, and safety culture.* NTSB symposium on corporate culture and transportation safety, Washington, DC.

Morey, J. C., Simon, R., Jay, G. D., Wears, R. L., Salisbury, M., Dukes, K. A., & Berns, S. D. (2002). Error reduction and performance improvement in the emergency department through formal teamwork training: Evaluation results of the MedTeams project. *Health Services Research, 37*(6), 1553–1581. doi:10.1111/hesr.2002.37.issue-6

Moriarity, D. (2015). *Practical human factors for pilots.* London, UK: Elsevier.

Naderpour, M., Lu, J., & Zhang, G. (2016). A safety-critical decision support system evaluation using situation awareness and workload measures. *Reliability Engineering & System Safety, 150*, 147–159. doi:10.1016/j.ress.2016.01.024

Northwest Airlines. (2005). *Flight operations manual (9.5.1)*. Eagan, MN: Northwest Airlines.

Preischl, W., & Hellmich, M. (2016). Human error probabilities from operational experience of German nuclear power plants, Part II. *Reliability Engineering & System Safety, 148*, 44–56. doi:10.1016/j.ress.2015.11.011

Salmon, P., Stanton, N., Walker, G., & Green, D. (2006). Situation awareness measurement: A review of applicability for C4i environments. *Applied Ergonomics, 37*(2), 225–238. doi:10.1016/j.apergo.2005.02.00

Savage, I. (2013). Comparing the fatality risks in United States transportation across modes and over time. *Research in Transportation Economics, 43*(1), 9–22. doi:10.1016/j.retrec.2012.12.011

Sigurdsson, S. O., Taylor, M. A., & Wirth, O. (2013). Discounting the value of safety: Effects of perceived risk and effort. *Journal of Safety Research, 46*, 127–134. doi:10.1016/j.jsr.2013.04.006

Simon, R., Langford, V., Locke, A., Morey, J. C., Risser, D., & Salisbury, M. (2000). A successful transfer of lessons learned in aviation psychology and flight safety to health care: the MedTeams system. *Patient Safety Initiative*, 45–49.

Skinner, B. F. (1957). *Verbal behavior*. New York, NY: Appleton-Century-Crofts.

Skinner, B. F. (1974). *About behaviorism*. New York, NY: Vintage.

Smith, G. S., Houmanfar, R., & Denny, M. (2012). Impact of rule accuracy on productivity and rumor in an organizational analog. *Journal of Organizational Behavior Management, 32*, 3–25. doi:10.1080/01608061.2012.646839

Smith, G. S., Houmanfar, R. A., & Louis, S. J. (2011). The participatory role of verbal behavior in an elaborated account of metacontingency: From conceptualization to investigation. *Behavior and Social Issues, 20*, 122–146.

Stewart, I., Barns-Holmes, D., Barnes-Holmes, Y., Bond, F. W., & Hayes, S. C. (2006). Relational frame theory and industrial/organizational psychology. *Journal of Organizational Behavior Management, 26*, 55–90. doi:10.1300/J075v26n01_03

Sullenberger, C., & Zaslow, J. (2009). *Highest duty: My search for what really matters*. New York, NY: William Morrow & Company.

Turner, P. (2012). Implementation of TeamSTEPPS in the emergency department. *Critical Care Nursing Quarterly, 35*(3), 208–212. doi:10.1097/CNQ.0b013e3182542c6c

United States Federal Aviation Administration (FAA). (1991). *Advisory circular: Aeronautical decision making*. AC No: 60-22. US Department of Transportation. Retrieved from http://rgl.faa.gov/Regulatory_and_Guidance_Library/rgAdvisoryCircular.nsf

United States Federal Aviation Administration (FAA). (2004). *Advisory circular: Crew resource management training*. Washington, DC: U.S. Dept. of Transportation, Federal Aviation Administration.

Vaughn, D. (1996). *The Challenger launch decision: Risky technology, culture, and deviance at NASA*. Chicago, IL: University of Chicago Press.

Vidulich, M. A., Stratton, M., Crabtree, M., & Wilson, G. (1994). Performance-based and physiological measures of situational awareness. *Aviation, Space, and Environmental Medicine, 65*(5), A7–A12.

Waag, W. L. (1994). Tools for assessing situational awareness in an operational fighter environment. *Aviation, Space, and Environmental Medicine, 65*(5), A13–A19.

Wagener, F., & Ison, D. (2014). Crew resource management in commercial aviation. *Journal of Aviation Technology and Engineering, 3*(2), 2–13. doi:10.7771/2159-6670.1077

Ward, M., Zhu, X., Lampman, M., & Stewart, G. (2015). TeamSTEPPS implementation in community hospitals. *International Journal of Quality Assurance, 28*(3), 234–244.

Wulfert, E., Greenway, D. E., Farkas, P., Hayes, S. C., & Dougher, S. C. (1994). Correlation between self-reported rigidity and rule-governed insensitivity to operant contingencies. *Journal of Applied Behavior Analysis, 27*(4), 659–671. doi:10.1901/jaba.1994.27-659

Yang, C. W., Yang, L. C., Cheng, T. C., Jou, Y. T., & Chiou, S.-W. (2012). Assessing mental workload and situation awareness in the evaluation of computerized procedures in the main control room. *Nuclear Engineering and Design, 250*, 713–719. doi:10.1016/j.nucengdes.2012.05.038

Zeltser, M. V., & Nash, D. B. (2010). Approaching the evidence basis for aviation-derived teamwork training in medicine. *American Journal of Medical Quality, 25*(1), 13–23. doi:10.1177/1062860609345664

Integrating Behavioral Science with Human Factors to Address Process Safety

Manuel A. Rodriguez, John Bell, Michelle Brown, and Donna Carter

ABSTRACT

Applying the science of human factors to eliminate error across all aspects of process design, management, operation, and maintenance has been a focus in the process safety area for many years. Human error has been attributed as a major cause of many high profile catastrophic accidents around the world. These accidents have resulted in national and international attention, which has led to a focus on improving organizational capabilities, systems, and in many cases, governmental regulations around human factors. This article provides a review of the field of human factors highlighting various topics in the literature, and introduces governmental regulatory bodies currently engaging organizations in a scientific approach to human factors. Finally, the need for integrating behavioral science methodologies with human factors is addressed. This is done with specific focus on how Organizational Behavior Management methodologies can work in concert with human factors to optimize process safety.

A massive explosion lifted the roof off of Reactor Building No. 4 at the Chernobyl Power Station, near Kiev nearly thirty years ago on the morning of April 26, 1986. The sequence of events that led to this horrific incident is well documented (Atherton & Frederic, 2008; Center for Chemical Process Safety/AIChE, 2011a; U.S. Nuclear Regulatory Commission, 2009) and continues to be covered by the media (Grunberg, 2014; Simon, 2014). The facility was attempting to conduct an experiment to improve the facility's response under emergency conditions. Fearing failure to perform the test because of some minor delay or procedural issues, the operating crew proceeded despite the absence of the technical staff or other limitations in the test procedure.

During the test, numerous operational policies were violated. When plant personnel made the decision to proceed with the test despite the instability of plant processes, the nuclear reaction accelerated rapidly, leaving the operators no time to regain control of the reactor (Center for Chemical Process Safety/AIChE, 2011a).

Human factors defined

The behaviors associated with the Chernobyl disaster and other major events are commonly referred to as human error. Human error is defined as any human action that exceeds some control limit as defined by the operating system (Center for Chemical Process Safety/AIChE, 2011a). The Health and Safety Executive (2005) have been able to classify human error under three overarching categories: unintentional failures or physical errors ("not doing what you meant to do"), mental errors where you do the wrong thing believing it to be right ("making the wrong decision"), and intentional failures or violations ("knowingly taking short cuts or not following known procedures").

A focus solely on human errors cannot provide the necessary vantage points to make a significant difference in process safety performance. Taking a systems-based approach to safety, including human behavior, is viewed more positively, and is the very essence of the field of human factors. The conceptual model of human factors is best summarized by the leading journal in the field, *Human Factors: The Journal of the Human Factors and Ergonomics Society* (Sage Publications, 2017). This scientific community publishes a peer-review journal on studies in human factors/ergonomics. The journal presents "theoretical and practical advances concerning the relationship between people and technologies, tools, environments, and systems" (Sage Publications, 2017). The science is described as "the basic understanding of cognitive, physical, behavioral, physiological, social, developmental, affective, and motivational aspects of human performance—to yield design principles; enhance training, selection, and communication; and ultimately improve human-system interfaces and sociotechnical systems that lead to safer and more effective outcomes" (Sage Publications, 2017). Before exploring the link Organizational Behavior Management (OBM) can and should have with human factors, a few more definitions of this field are worth noting.

The Center for Chemical Process Safety/AIChE (2003) defines human factors as the scientific discipline of understanding how human interaction with a system influences the behavior of the person and the performance of the system, as well as applying theory and data to optimize the performance of both. The second definition offered by the Health and Safety Executive (2015a) states human factors refer to the environmental, organizational, job, and individual characteristics that influence our behavior in ways that can impact health and safety. Both definitions lend themselves for study, application, and theorizing possible alternatives or integrations with other disciplines, such as OBM.

Human factors also considers an organizational level of analysis that includes work patterns, the culture of the workplace, resources, communications, and leadership. When jobs are designed these organizational factors are often overlooked although they can significantly influence behavior (Health and Safety Executive, 2015a). The Center for Chemical Process Safety/AIChE (2011b) supports this stance by placing emphasis on how an organization's safety culture, shared values and behaviors, can greatly influence the manner in which process safety is managed. They suggest that cultural root causes should increasingly be considered when analyzing process safety performance problems.

Although human error is often recognized as being a contributor to incidents and accidents, very few organizations are able to proactively manage this (Health and Safety Executive, 2005). Table 1 provides three levels of proactive management to human factors provided by the Health and Safety Executive, and is part of the inspector's toolkit for evaluating human factors in major hazard facilities. In the majority of investigations of catastrophic incidents such as the accident at the Chernobyl nuclear power plant in 1986 (World Nuclear Association, 2017), and in 2005, a series of explosions that occurred at the BP Texas City refinery during the restarting of a hydrocarbon isomerization unit (U.S. Chemical Safety and Hazard Investigation Board, 2007) are two of the many incidents, which illustrate that common process safety culture weaknesses were often identified listing (a) the lack of behavioral contingencies such as expectations and enforcement of process safety standards, (b) process safety controls were primarily "check-the-box" activities, and (c) the lack of feedback systems that initiated timely responses to process safety issues and concerns (Center for Chemical Process Safety/AIChE, 2011b). Overt links to critical behaviors of omission (i.e., neglecting to perform a behavior related to Process Safety Management) or commission (i.e., performing behaviors that conflict with, or hinder the controls attempted by Process Safety Management) tend to be associated with catastrophes such as Chernobyl and Texas City.

Table 1. From the Inspector's Toolkit: Human Factors in Major Hazards (Health and Safety Executive, 2005).

Level 1: Core topics	
1.1 Competence assurance	Fundamental to good human factors arrangements at all sites
1.2 Human factors in accident investigation	
1.3 Identifying human failure	
1.4 Reliability and usability of procedures	
Level 2: Common topics	
2.1 Emergency response	Relevant human factors subjects at most sites
2.2 Maintenance error	
2.3 Safety critical communications	
2.4 Safety culture	
Level 3: Specific topics	
3.1 Alarm handling and control room design	Important human factors issues but only for some sites at some times
3.2 Managing fatigue risks	
3.3 Organizational change and transition management	

Human factors and OBM

The contribution of the Health and Safety Executive (2015a) list of focus areas illustrates the complexity of process safety management and how many factors can influence a person's performance within a system:

(1) risk assessment and incident investigation;
(2) procedures;
(3) training and competence;
(4) staffing levels;
(5) workload;
(6) supervision;
(7) contractor management;
(8) organizational change;
(9) safety critical communications, including shift handover and permit to work;
(10) human factors in design (control rooms, human-computer interfaces (HCI), alarm management, and lighting, thermal comfort, noise and vibration);
(11) fatigue management and shiftwork;
(12) organizational culture (behavioral safety and learning organizations); and
(13) maintenance, inspection and testing.

The Health and Safety Executive have demonstrated that the success of any process safety management system lies in an organization's ability to predict and prevent deviations from set standards, all of which involve performance and maintenance of critical safety behaviors. Integration of human factors systems with behavioral science can ensure critical process safety behaviors are consistently performed.

The next section will focus on four behavioral science approaches to human performance improvement and will argue for the utility of an integrated approach between behavioral science and the area of human factors as related to the 13 topics described by the Health and Safety Executive (2015a). The discussion in this article is conceptual in nature, although some examples come from direct applications. The approach suggested is intended to encourage practitioners and researchers to understand OBM methodologies that can aid in the advancement of human factors, aimed at a generalizable approach as human factors is applied across a number of high tech, high integrity operations (chemical, oil and gas exploration, aviation, transportation, nuclear, medicine, etc.). These industries face similar challenges as automation, instrumentation, and multicultural workforces interlock in complex, dynamic operations.

Behavioral systems analysis

Behavioral Systems Analysis (BSA) focuses on performance improvement in organizations through the methods and principles of behavior analysis and systems analysis (Brethower, 2015). More specifically, BSA is the design and modification of systems that lead to overall improvement of an organization (Malott & Garcia, 1987). By identifying and understanding all of the system components within an organization, management teams can work to improve any inconsistencies or conflicting requirements, resulting in a system where all components produce outputs for the same objective. Brethower (2015) observed BSA as being noticeably effective at increasing productivity, improving safety performance, reducing quality cycle time and improving customer service across many organizations.

The literature illustrates a clear opportunity for BSA integration with the existing human factors theories and methodologies. Through the lens of human factors, the organization is viewed in the layers of organization, job, and individual (Health and Safety Executive, 2015a). This structure is replicated in BSA with the analyses conducted at the organization, process, and performer levels (Diener, McGee, & Miguel, 2009). Each business unit is looked upon as a Total Performance System, within which exists contingencies associated with the purpose of the work, outputs, inputs, feedback from external parties or stakeholders, processes, internal feedback, the external operating environment, and external/internal competition.

Completing this analysis across the three layers allows managers to see where they have competing contingencies, which result in the omission of critical process safety behaviors (Diener et al., 2009; Health and Safety Executive, 2015a). BSA assessment tools such as the Behavior Systems Analysis Questionnaire (BSAQ) (Diener et al., 2009) could be integrated with the existing human factors systems model and utilized by process safety professionals looking to ensure a comprehensive evaluation of human factors is conducted prior to implementing any change strategies. The BSAQ (2009) provides critical questions that could support a human factors study such as "does the organization have sufficient human resources? What Materials and Equipment are needed? Where do they come from? What information is collected about processing system performance? (e.g., quality, quantity, timeliness, cost, safety); How is this information measured and used? By Who?"

Behavior that supports process safety management will only exist if there are environmental contingencies in place to support it. BSA recognizes different contingencies exist at different levels of an organization, and therefore require different forms of intervention and, more important, alignment of these contingencies through organizational layers (Diener et al., 2009). Examples of the varying differences of contingencies can be highlighted from the board room where executives make decisions based on stakeholder input and market demand, to midlevel management being influenced by both upper and lower levels of an organization to make decisions and/or carry out strategic priorities, a term illustrating this phenomenon more commonly referred to in the OBM literature as metacontingencies (Mawhinney, 2001), and where employees work day-in-day-out, contingencies are typically shaped by the social environment of peer to peer interactions, and the very antecedents and consequences intended to influence employees behaviors (e.g., rules, task clarification, and feedback)

The foundations of BSA and human factors are similar with regard to how they view the impact of the organization on the performer and vice versa. Inclusion of BSA into existing human factors systems would aid organizations in understanding why particular behaviors are often found to be contributory in serious process incidents. The methodical, multilayer, multidirectional analysis of contingencies that impact the organization allows for more precise management of system factors impacting the performer.

Three of the current authors conducted a behavioral systems approach that was taken during a process safety event at a chemical manufacturing facility in North America. Legal constraints do not permit the authors to disclose a detailed account of the event, however, here is a brief summary. The plant experienced a process incident leading to a chemical fire. The plant operators assessed the situation and quickly followed the emergency response protocol, which included a request for the local fire department to respond. All chemicals were fully contained, fire damage was substantial, one employee died from a failed attempt to evacuate, and news coverage of this event occurred immediately due to coverage through social media before the organization headquarters even knew of the incident. This incident was investigated by the organization, the local authorities, and government officials. Findings from the investigation concluded a systemic gap around job level factors such as confusing directives on evacuation procedures, procedural factors such as inadequate knowledge transfer and inadequate technical designs from an engineering standpoint, and organizational factors such as inadequate horizontal and vertical communications and inadequate standards compared to benchmarking studies performed in previous years. The BSA uncovered various levels of contingencies and metacontingencies that could have been modified to prevent such an incident.

Behavior-based safety

Behavior-Based Safety (BBS) is defined as "focusing on the analysis and alteration of work environments to reduce injuries and promote safe behavior among leaders and employees" (Organizational Behavior Management Network, 2015a). This evidence-based approach has traditionally focused on identifying safe and at-risk behaviors indigenous to the workplace and then engaging in direct observation of the behaviors followed by direct feedback.

Mature BBS processes then collect the data from observations to engage in behavioral functional analyses to design and implement environmental and reinforcement interventions to promote and sustain safe behavior. Although BBS has been associated with substantial reduction of injuries (Sulzer-Azaroff & Austin, 2000) the integration of BBS within process safety efforts has seen very little research (Hyten, 2010). Anderson (2005) supports the use of BBS in accident prevention, however, suggests to readers there needs to be a balanced approach to tailoring BBS programs to the special topics in process safety. To do so we may need to progress BBS beyond on-task safe behaviors to include or substitute the types of safety management behaviors integral to process safety so they can receive the level of observation and analysis they require.

Of the thirteen topics from the Health and Safety Executive (2015a), procedural use, shift handover, permit to work, HCI, and maintenance inspections may benefit from a BBS approach to improve process safety and specifically human factors. Each of these areas include behaviors, which occur at high frequency, that are susceptible to human error, and require understanding of the contingencies, which enable the behaviors to occur at desired levels. Contingencies to be examined should include the role of instruction, rules and whether rule-governed behavior is occurring (Malott, 1993; Maraccini, Houmanfar, & Szarko, 2016). Options for integration could include addition of human factors safety management behaviors to a task observation card, a task specific ABC analysis to determine what contingencies are influencing human factors safety management behaviors, and peer feedback on performance of these specific behaviors. If applied to maintenance inspections, a BBS review of the procedure would determine whether or not the behaviors associated with quality inspections have been identified and if the task has been designed to promote performance of the inspection behaviors. Observations could be completed, which would encourage correct inspection behaviors and would provide opportunity to correct undesired behaviors.

By taking a BBS approach to these focus areas, practitioners and researchers can begin to further develop refined methods of improving behavior systematically, ultimately improving process safety performance.

Fluency-based training and development

The area of training and competence is highly regarded as critical to safe operations and human factors interventions. The Health and Safety Executive (2015a) call on training programs so that the job incumbent can regularly perform their tasks to a recognized standard and have a proficient understanding of the responsibilities associated with the task. Integration with OBM would see the introduction of fluency-based training to organizations' task-based training.

Fluency training differs from traditional methods of task-based training because it focuses on training performers to a level, which allows them to perform quickly, accurately, and without hesitation (Binder, 1996). In the opening to his article *Behavioral Fluency: Evolution of a New Paradigm*, Carl Binder (1996, p. 163) wrote "fluency-based education and training programs have produced some of the most dramatic results in the history of behaviorally oriented instruction." Training to fluency ensures the mastery of basic foundational skills, making more advanced work easier rather than harder. Basic foundational skills are most commonly improved upon by increasing the accuracy and frequency of the desired response. Published and unpublished research has demonstrated that when desired frequencies of accurate performance are achieved, learners seem to retain and maintain what they have learned, remain on-task in the face of distraction, and generalize those performances to new situations (Binder, 1996). For example, Pampino, Wilder, and Binder (2005) utilized instructional and measurement procedures based on precision teaching designed to increase rates of accurate responding of product knowledge and data entry skills in a large construction organization.

There are synergies between how OBM literature describes fluency-based training and the eleven key principles in competence described by the Health and Safety Executive (2015a). As the Health and Safety Executive (2015a) define it, competence "is the ability to undertake responsibilities and perform activities to a recognized standard on a regular basis. It is a combination of skills, experience and knowledge." Focusing on a "standard" of performance and the occurrence of the performance "on a regular basis" is parallel to work in OBM on fluency-based performance, focusing on a defined set of behaviors to build fluency against some criteria and then measuring the behaviors based on accuracy, repeatability, and timeliness of response to a given stimulus. Table 2 provides a summary side-by-side comparison of the key principles in competence from the Health and Safety Executive and literature on fluency-based training. The study and application of fluency-based training and development can enable human factors by accelerating the performance in safety training and competency programs. One could argue that taking a fluency-based training and development approach to safety in the area of human factors could be a paradigm shift, making way for a series of positive improvements, including reduced injury rates, and reliable process safety.

Leadership and culture

Krapfl and Kruja (2015, p. 36) describe organizational culture as "the way we do things around here"; a definition shared by the Health and Safety Executive (2015a). The Health and Safety Executive continued by describing

Table 2. Illustration Between Health and Safety (2015a) Key Principles in Competence and Fluency-Based Training Literature.

Health and Safety (2015a) Key principles in competence	Literature on fluency-based training	Citation
1. Competence assurance should be linked to key responsibilities, activities, and tasks identified in risk assessments.	• Select the precise skill or behavior that should be targeted based on this assessment. • Fluency training focuses on high performance aims and custom-tailored program to maximize learning.	• NEPS (2012) Lindsley (1992)
2. Competency assurance systems should aim to establish and maintain competency for all those involved in safety-related work, including managers.	• Training should ensure target skills are maintained long-term, should generalize, and help to avoid inappropriate approximations of the skill as the learner progresses • Fluency-based training incorporates all of the stages of learning.	• NEPS (2012); Lindsley (1992) Binder (1996)
3. Training is an important component of establishing competency but is not sufficient on its own.	• A learner needs to achieve a level of fluency • Fluency linked to improved retention.	• White (1986); Johnston & Pennypacker (1980); Bucklin, Dickinson, and Brethower (2000) Bucklin et al. (2000)
4. Competence assurance systems should take account of foreseeable work and operating conditions—including infrequent and complex activities, emergency situations and upsets, maintenance etc.	• Structure the learner's environment and/or task to allow for success and improvement. If success is not seen, alter the environment and/or task.	• NEPS (2012)
5. Training and competence assessment methods should be appropriate to the hazard profile of the tasks being undertaken.	• Not all skills are the same level of difficulty; some may be easier to acquire than others. • More difficult skills need to be practiced more often, but for multiple short periods of time to ensure accuracy and fluency.	• NEPS (2012) Binder, Haughton, & Eyk (1990)
6. "On-the-job" training should be structured and linked to risk assessments and associated control measures including procedures. In safety critical environments, on-the-job training should be supported by other forms of training where appropriate.	• Fluency-based training involves multiple tools, tactics, and simplified methods for maximum learning. These tools and methods comprise the body of precision/fluency-based training. PT improves or combines well with any curriculum.	• Lindsley (1992)
7. Training should be validated ("Did it deliver what it was supposed to?"), evaluated ("Is this the right kind of training for our needs?") and recorded.	• Over successive days of practice, chart data and monitor progress to determine whether the task is too easy, too difficult, or just right to ensure accurate knowledge of the skills being trained. Was this skill acquired? Is this skill appropriate for the specific learner's needs?	• NEPS (2012)

(*Continued*)

Table 2. (Continued).

Health and Safety (2015a) Key principles in competence	Literature on fluency-based training	Citation
8. There should be refresher training for infrequent, complex, or safety critical tasks and this may include appropriate reassessment.	• Re-evaluate the skill being taught if the learner acquires the skills too quickly or is not successful	• NEPS (2012); Lindsley (1992); White (1986)
9. Vocational qualifications should include site-specific aspects and link appropriately to the hazards and risks in your workplace.	• See #4 and #5 for possible links that may apply here.	
10. Aim to achieve a suitable balance between competence and supervision.	• Monitoring the learner's progress and performance of a targeted skill should be frequent and direct in the beginning; once progress is consistent and/or reached the designated level of improvement—you can fade out and move on to the next complex skill.	• Sugai and Horner (2006)
11. Careful consideration should be given to the potential consequences of outsourcing of safety-related work.	• See #1 through 7 for possible links that may apply here.	

how organizational culture "influences human behavior and human performance at work." The influences listed are characteristics such as management commitment and style, employee involvement, communication, and organizational learning. These categories are related to critical management behaviors, which interlock with on-the-ground behaviors related to process safety (Ludwig, 2017).

Leadership and culture is not a new topic in the field of OBM. Several examples of OBM methodologies applied to leadership and culture have been published in a wide range of communication mediums (Baum, 1994; Beers, 1998, Braksick, 2007; Callahan, 2000; Daniels, 1994; Daniels & Daniels, 2006; Daniels and Bailey, 2014; Eubanks & Lloyd, 1992; Komaki, 1998; Mawhinney, 1992b, 2006, 1992a; Mawhinney & Ford, 1977; Rao & Mawhinney, 1991; Redmon & Mason, 2001; Rodriguez, Neff, Robertson, Kelley, & King, 2015; Ventura, Vazquez, & Rodriguez, 2015; Wilson, 1999). In 2015 the *Journal of Organizational Behavior Management* published a special issue on Leadership and Cultural Change, focusing on how behavior analysis can support investigating the behavior of leaders (Houmanfar & Mattaini, 2015). With the collection of literature on the subject matter on leadership and culture, OBM is positioned to enable human factors to have a positive impact on leadership behavior toward process safety objectives.

OBM can further enhance leadership capabilities by using scientific-based studies, which identify behaviors often exhibited by effective leaders. Krapfl and Kruja (2015) provide a "menu" of leadership behaviors that could support the Health and Safety Executive's (2015b) tool set:

- evaluating the value proposition;
- behaving ethically and with integrity by demonstrating commitment to all of the organizational constituencies involved;
- execution skills (making things happen);
- entrepreneurial and innovation skills (thinking outside of the box);
- communication skills, which includes sharing information clearly and listening;
- enabling skills;
- team building skills;
- confronting adversity;
- tenacity; and
- culture-building skills.

The last behavior class (culture-building skills) reinforces the role leaders play in influencing culture-building skills. Similarly, decision making has been researched in the field of OBM (Bumstead & Boyce, 2005; Hantula & Crowell, 1994; Rawlins, 1990), thus a behavioral analytical approach to decision making could be integrated into a more complete treatment of process safety in human factors.

Human factors also identifies behavior classes, which leaders need to meet in order for process safety to be effectively managed in their teams (Health and Safety Executive, 2015b). While these competencies have been derived from incident data and hold statistical validity, their effectiveness is limited without the addition of OBM methodologies. Using antecedents such as training leaders to meet specific competencies and understanding the broader environmental contingencies involved may result in minimizing or eliminating deficits in the managerial behaviors needed to set the contingencies for their workforce behaviors to preserve process safety.

Conclusion

Human factors seeks to integrate human dispositions into job-level designs. Similarly, the Health and Safety Executive (2015a) includes cognitive dispositions such as perception, attention, and decision making while engaging in tasks related to process safety. OBM methodologies such as BBS and fluency-based training can supplement and potentially strengthen desired human factors focus areas at the job and individual level by interpreting cognitive constructs such as competence, skills, personality, attitude, job enrichment, equipment and processes, and risk perception into behaviorally-sound actionable operations. In addition, OBM methodologies such as BSA and the behavioral focus on leadership and culture can enhance organizational and management level focus areas in regard to human factors such as work planning, safety systems, incident response efforts, management communications, and health and safety culture.

Human factors has often referred to as "the thread that runs through the safety management system, the organization for safety, and the culture of a site"(Health and Safety Executive, 2005, p. 7). Many practitioners in OBM would suggest behavior or the "thread" that is, human factors is often misunderstood, mismanaged, or even simply missed all together. Because human behavior is often identified as the root cause of incidents leading to process safety events, individuals who lead or participate in investigations need to further their knowledge and skills in the understanding of human behavior in regard to the environment, contingencies, and effective behavior management techniques.

Because technology continues to evolve faster than the ability to predict how humans will interact with it, industries such as manufacturing and aeronautics can no longer depend as much on experience and intuition to guide decisions related to human factors. Instead, a sound scientific basis is necessary for assessing human behavior to make a positive difference across the various human factors topics described in this article. Meanwhile, unless the field of OBM takes strides toward integration with such disciplines as human factors, organizations will continue to make major investments in training, equipment, and systems that may have a short-term impact, but may fall short, resulting in major process failures, equipment damage, or, worse yet, personal injury and loss of life.

References

Anderson, M. (2005). Behavioral safety and major accident hazards: Magic bullet or shot in the dark? Process Safety and Environmental Protection, *83*(2), 109–116.

Atherton, J., & Frederic, G. (2008). *Incidents that define process safety. Center for chemical process safety of the American Institute of Chemical Engineers*. Hoboken, NJ: John Wiley & Sons, Inc.

Baum, W. M. (1994). *Understanding behaviorism: Science, behavior, and culture*. New York, NY: HarperCollins.

Beers, J. (1998). Shaping as a leadership tool for sustaining behavior change. In B. Hopkins (Ed.), *Sustaining organizational performance improvement for the long term*. Symposium conducted at the 24th annual convention of the Association for Behavior Analysis, Orlando, FL. Association for Behavior Analysis.

Binder, C. (1996). Behavioral fluency: Evolution of a new paradigm. *The Behavior Analyst*, *19*, 163–197.

Binder, C., Haughton, E., & Van Eyk, D. (1990). Increasing endurance by building fluency: Precision Teaching attention span. *Teaching Exceptional Children*, *22*, 24–27. doi:10.1177/004005999002200305

Braksick, L. W. (2007). *Unlock behavior, unleash profits: Developing leadership behavior that drives profitability in your organization*. New York, NY: McGraw-Hill Companies.

Brethower, D. M. (2015). *Behavioral systems analysis: What is behavioral systems analysis?* Retrieved September 11, 2015 from http://www.behavior.org/resource.php?id=395

Bucklin, B. R., Dickinson, A. M., & Brethower, D. M. (2000). A comparison of the effects of fluency training and accuracy training on application and retention. *Performance Improvement Quarterly*, *13*(3), 140–163. doi:10.1111/j.1937-8327.2000.tb00180.x

Bumstead, A., & Boyce, T. E. (2005). Exploring the effects of cultural variables in the implementation of behavior-based safety in two organizations. *Journal of Organizational Behavior Management*, *24*(4), 43–63. doi:10.1300/J075v24n04_03

Callahan, D. (2000). *Implementing reinforcement based leadership at Chevron Chemical Company*. Paper presented at the Association for Behavior Analysis meetings, Washington, DC.

Center for Chemical Process Safety/AIChE. (2007). *Human factors methods for improving performance in the process industries*. Hoboken, NJ: John Wiley & Sons, Inc..

Center for Chemical Process Safety/AIChE. (2011a). The importance of human factors. In *Conduct of operations and operational discipline--For improving process safety in industry* (Chap. 4). Hoboken, NJ: John Wiley & Sons, Inc.

Center for Chemical Process Safety/AIChE. (2011b). Process safety culture. In *Guidelines for auditing process safety management systems* (2nd ed.) (Chap. 4). Hoboken, NJ: John Wiley & Sons, Inc..

Daniels, A. C., & Daniels, J. E. (2007). *Measure of a leader: The legendary leadership formula for producing exceptional performers and outstanding results*. Atlanta, GA: McGraw-Hill Education.

Daniels, A. C., & Bailey, J. S. (2014, April 15). *Performance management: Changing behavior that drives organizational performance* (5th ed.). Atlanta, GA: Performance Management Publications.

Daniels, A. C. & Daniels, J. E. (2005). *Measures of a leader*. Tucker, GA: Performance Management Publications.

Diener, L. H., McGee, H. M., & Miguel, C. F. (2009). An integrated approach for conducting a behavioral systems analysis. *Journal of Organizational Behavior Management*, *29*(2), 108–135. doi:10.1080/01608060902874534

Eubanks, J. L., & Lloyd, K. E. (1992). Relating behavior analysis to the organizational concept and perspective. In T. C. Mawhinney (Ed.), *Organizational culture, rule governed behavior and organizational behavior management: Theoretical foundations and implications for research and practice*. New York, NY: Haworth Press.

Geller, E. S. (2001). *Working safe: How to help people actively care for health and safety* (2nd ed.). Boca Raton, FL: CRC Press.

Hantula, D. A., & Crowell, C. R. (1994). Intermittent reinforcement and escalation processes in sequential decision making: A replication and theoretical analysis. *Journal of Organizational Behavior Management*, *14*(2), 7–36. doi:10.1300/J075v14n02_03

Health and Safety Executive. (2005). *Human factors: Inspectors human factors toolkit*. Retrieved August 23, 2015, from http://www.hse.gov.uk/humanfactors/toolkit.htm

Health and Safety Executive. (2015a). *Human factors: Managing human performance - Briefing notes*. Retrieved August 23, 2015, from http://www.hse.gov.uk/humanfactors/index.htm

Health and Safety Executive. (2015b). *Leading health and safety at work*. Retrieved August 23, 2015, from http://www.hse.gov.uk/leadership/index.htm

Houmanfar, R. A., & Mattaini, M. A. (2015). Leadership and cultural change. *Journal of Organizational Behavior Management, 35*(1–2), 1–3. doi:10.1080/01608061.2015.1036645

Hyten, C. (2010). *Integrating strengths of behavior-based safety with effective process safety management*. Retrieved September 11, 2015, from http://aubreydaniels.com/pmezine/integrating-strengths-behavior-based-safety-effective-process-safety-management

Johnston, J. M., & Pennypacker, H. S. (1980). *Strategies and tactics of human behavioural research*. Hillsdale, NJ: Lawrence Erlbaum Associates.

Komaki, J. L. (1998). *Leadership from an operant perspective*. New York, NY: Routledge.

Krapfl, J. E., & Kruja, B. (2015). Leadership and culture. *Journal of Organizational Behavior Management, 35*(1–2), 28–43. doi:10.1080/01608061.2015.1031431

Lindsley, O. R. (1992). Precision teaching: Discoveries and effects. *Journal of Applied Behavior Analysis, 25*, 51–57. doi:10.1901/jaba.1992.25-51

Ludwig, T. (2017). Process safety behavioral systems: Behaviors interlock in complex process safety meta-contingencies. *Journal of Organizational Behavior Management, 37*(3–4), 224–239.

Malott, R. W. (1993). A Theory of rule-governed behavior and organizational behavior management. *Journal of Organizational Behavior Management, 12*(2), 45–65. Doi:10.1300/J075v12n02_03

Malott, R. W., & Garcia, M. E. (1988). A goal-directed model for the design of human performance systems. *Journal of Organizational Behavior Management, 9*(1), 125–159.

Maraccini, A. M., Houmanfar, R. A., & Szarko, A. J. (2016). Motivation and complex verbal phenomena: Implications for organizational research and practice. *Journal Of Organizational Behavior Management, 36*(4), 282–300. Doi:10.1080/01608061.2016.1211062

Mawhinney, T. C. (Ed.). (1992a). *Organizational culture, rule-governed behavior and organizational behavior management: Theoretical foundations and implications for research and Practice*. New York, NY: The Haworth Press, Inc.

Mawhinney, T. C. (1992b). Evolution of organizational cultures as selection by consequences: The Gaia hypothesis, metacontingencies, and organizational ecology. *Journal of Organizational Behavior Management, 12*(2), 1–26.

Mawhinney, T. C. (2001). Organization-environment systems as OBM intervention context: Minding your metacontingencies. In L. J. Hayes, J. Austin, R. Houmanfar, & M. C. Clayton (Eds.), *Organizational change* (pp. 137–165). Reno, NV: Context.

Mawhinney, T. C. (2006). Effective leadership in superior-subordinate dyads. *Journal of Organizational Behavior Management, 25*(4), 37–79. doi:10.1300/J075v25n04_03

Mawhinney, T. C., & Ford, J. D. (1977). The path goal theory of leader effectiveness: An operant interpretation. *Academy of Management Review, 2*, 398–411.

National Educational Psychological Service. (2012). *Teaching site vocabulary and improving reading fluency: A good practice guide for teachers and parents*. Department of Education and Skills. Retrieved from http://www.education.ie/en/Publications/Media-Library/Literacy-Resources/NEPS-Resources.html

Organizational Behavior Management Network. (2015). *About organizational behavior management (OBM)*. Retrieved September 10, 2015, from http://www.obmnetwork.com/what_is_obm/definition_description_common_applications

Pampino, R. N., Jr., Wilder, D. A., & Binder, C. (2005). The use of functional assessment and frequency building procedures to increase product knowledge and data entry skills among foremen in a construction organization. *Journal of Organizational Behavior Management, 25*(2), 1–36. doi:10.1300/J075v25n02_01

Rao, R. K., & Mawhinney, T. C. (1991). Superior-subordinate dyads: Dependence of leader effectiveness on mutual reinforcement contingencies. *Journal of the Experimental Analysis of Behavior, 56*, 105–118. doi:10.1901/jeab.1991.56-105

Rawlins, C. (1990). The impact of teleconferencing on the leadership of small decision-making groups. *Journal of Organizational Behavior Management, 14*(2), 7–36.

Redmon, W. K., & Mason, M. (2001). Organizational culture and behavioral systems analysis. In C. M. Johnson, W. K. Redmon, & T. C. Mawhinney (Eds.), *Handbook of performance management: Behavior and management*. New York, NY: The Haworth Press, Inc.

Rodriguez, M., Neff, B., Robertson, C., Kelley, D., & King, A. (2015). *OBM applications in human services, leadership and culture*. Symposium conducted at the 35th annual convention of the Florida Association for Behavior Analysis, Daytona Beach, Florida.

Sage Publishing. (2017). *Human factors: The journal of the human factors and ergonomics society*. Retrieved March 17, 2017, from https://us.sagepub.com/en-us/nam/human-factors/journal201912#description

Simon, B. (2014). *Chernobyl: The catastrophe that never ended*. CBS News. Retrieved from http://www.cbsnews.com/news/chernobyl-the-catastrophe-that-never-ended/

Sugai, G., & Horner, R. R. (2006). A promising approach for expanding and sustaining school-wide positive behavior support. *School Psychology Review, 35*, 245.

Sulzer-Azaroff, B., & Austin, J. (2000). Does BBS work? Behavior-based safety & injury reduction: A survey of the evidence. *American Society of Safety Engineers, July 2000*, 19–24.

U.S. Chemical Safety and Hazard Investigation Board. (2007, March). *Investigation report: Refinery explosion and fire (15 killed, 180 injured). BP, Texas City, Texas, March 23, 2005* (Report No. 2005-04-I-TX). Retrieved March 17, 2017, from file:///Users/mannyrodriguez/Downloads/CSBFinalReportBP.pdf

U.S. Nuclear Regulatory Commission. (2009, April). *Background on Chernobyl nuclear power plant accident*. Washington, DC: Author.

Ventura, A., Vazquez, J., & Rodriguez, M. (2015). *Growing a behavioral organization ethically utilizing behavioral systems analysis and leadership*. Symposium conducted at the 35th annual convention of the Florida Association for Behavior Analysis, Daytona Beach, FL.

White, O. R. (1986). Precision teaching—Precision learning. *Exceptional Children, 52*, 522–534. doi:10.1177/001440298605200605

Wilson, N. (1999). Extending learnings from reinforcement based leadership to the overseas business. In T. Thurkow (Ed.), *Shaping Chevron Chemical Company's success through behavioral and systems changes*. Symposium conducted at the 25th annual convention of the Association for Behavior Analysis. Chicago, IL.

World Nuclear Association. (2017). *Chernobyl accident 1986, updated November 2016*. Retrieved March 17, 2017, from http://www.world-nuclear.org/information-library/safety-and-security/safety-of-plants/chernobyl-accident.aspx

Leadership's Role in Process Safety: An Understanding of Behavioral Science Among Managers and Executives Is Needed

Nicole Gravina, Bob Cummins, and John Austin

ABSTRACT
A recent prevalence of high visibility catastrophic events has garnered increased attention to process safety issues. While the use of Behavior-Based Safety interventions demonstrate a reduction in workplace injuries by targeting employee behavior, we believe that process safety requires a greater focus on the behavior of leaders (e.g., creating and executing strategy). One effective method to begin targeting leader behavior for the improvement of process safety is to teach leaders about the principles of behavior, including ways by which the science may be applied within their own organizational models.

Safety-related incidents and injuries have been on the decline due in part to an increased focus and greater understanding of effective prevention strategies (Bureau of Labor Statistics, 2014). More recently, process safety has garnered further attention as researchers and practitioners learn that the strategies used to reduce and prevent typical workplace illnesses and injuries may not be the same strategies required to prevent catastrophic events (OSHA, 2000). In fact, while workplace injuries have steadily declined over the past several years, fatality rates have remained unchanged (Bureau of Labor Statistics, 2014). This dispels the notion that tactics aimed at reducing less serious injuries will result in a reduction of more serious injuries, as some imply from Heinrich's Safety Triangle (Behavioral Safety Technologies, 2011; McSween & Moran, 2017).

Behavior-Based Safety (BBS) has demonstrated success for improving worker safety behaviors and reducing injuries (Tuncel, Lotlikar, Salem & Daraiseh, 2006). BBS typically involves measuring worker safety behaviors through peer observations or self-monitoring and providing feedback, both immediately to the individual and aggregately to the group, in order to improve safety performance (McSween, 2003). The safety behaviors are identified through an assessment that involves examining injury and close call data as well as observation and input from employees and supervisors. It tends to focus on regularly occurring worker behaviors that can be frequently measured and coached. Although it is effective for improving workplace safety, the approach may not be adequate for preventing more serious or catastrophic events tied to behaviors that occur less frequently.

Many process safety incidents occur during nonroutine tasks, when unexpected events arise and employees must immediately respond or when equipment that is mostly run nonstop is being shut down, restarted, or fixed. For example, an explosion occurred at both a wastewater site in Texas in 1990 and an oilfield in Mississippi in 2006 during a non-standard maintenance task, and another at a fertilizer plant in Iowa in 1994 after it had been shut down (Ness, 2015). This suggests that new interventions that can help eliminate safety incidents related to behaviors outside of the typical BBS observation checklist need to be developed and tested.

Programs like BBS tend to be managed almost entirely by frontline employees and supervisors, failing to engage important influencers of safety who are found among managers and executives. It is widely recognized in the health and safety literature that leadership plays an important role in health and safety (e.g., Zohar, 2002), but with a few exceptions (e.g., Komaki & Citera, 1990; Komaki, Desselles, & Bowman, 1989), leadership behavior has received little empirical attention in Organizational Behavior Management (OBM). Consultants have attempted to use behavior checklists to measure critical leadership behavior (e.g., accompanying an employee during an observation, addressing a safety barrier), in the same way checklists are used for employee behavior (Cooper, 2006). However, leadership behavior tends to be much more heuristic and dynamic because leaders are required to problem solve and make decisions at high levels that effect safety directly or indirectly on a regular basis (Smith, Lewis, & Tushman, 2016), which makes it difficult to observe with checklists. For example, budget decisions can directly or indirectly impact safety in a number of ways, both immediately and in the future. In addition, managers and executives' peers may be too busy or dispersed to conduct observations amongs themselves and their direct reports may not feel comfortable providing feedback. Nonetheless, managers and executives engage in behaviors that impact safety and those behaviors are subject to the principles of behavior in the same way as worker behaviors.

The work culture and environment are influenced by manager and executive behavior, including verbal behavior, on a daily basis. This influence can be haphazard or by purposeful design. The work environment consists of both the physical structures (e.g., equipment, checklists) and the actions and interactions among employees at all levels of the business (Ludwig, 2017). Leaders manage many aspects of work that are out of the control of workers and frontline supervisors. This means workers and supervisors work in an environment that has been mostly created by the managers and executives.

Even if frontline employees are working at their best, they may not be able to prevent a catastrophic event without the influence and strategic decision making from leaders that support safety. Before considering ways to influence leader behaviors to impact process safety, it first must be understood how leader behavior impacts process safety.

Linking leadership behavior to process safety issues

A paper by Bell and Healey (2006) reviewed case studies of process safety incidents, identified causes, and grouped them into categories. The causes they identified revealed that management and leadership practices play an important role in process safety. Some of the contributing variables were directly linked to leadership behavior, like ineffective or insufficient management practices, poor systems for managing safety, and communication issues between employees and leaders. Others included situations created by managers, such as setting difficult-to-meet goals for production, long work hours, and changing equipment or procedures without appropriately informing or training the worker. Further contributing factors to process safety events were related to issues that managers have direct influence over and could detect and improve, for instance, insufficient or ineffective systems for reporting issues and concerns, problems with maintenance work and completion, poor systems to train and coach employee knowledge and skills, failure to adjust and improve after incidents had occurred, and workers not following correct procedures. The authors acknowledged that more work is needed to hone in on causes of process safety issues but from their review it is clear that the environment created by managers and executives plays a significant role in process safety incidents.

Each of these causes pinpointed by Bell and Healey encompasses a collection of leader behavior, many of them involving verbal behavior (e.g., communication with employees) and decision making. For example, pressure to meet production targets may be a result of intentional or inadvertent verbal behavior during meetings specifying stimuli such as targets, difficult to attain goals, or reactions emitted when targets have not been met. The authors of the current paper have encountered a number of coaching situations when a leader made a statement they considered relatively benign (e.g., "We need to meet our production targets by the end of the week or we may have to work through the weekend") that resulted in unintended behaviors from the workers hearing the message (e.g., taking shortcuts and working longer hours during the week).

In addition, we have encountered many situations where leaders made a decision without fully understanding the downstream impact of that decision. For example, an organization rolled out a "fix-it-myself" initiative encouraging employees to try to fix equipment themselves before bringing the issue to maintenance and including this as a point of evaluation on performance appraisals. This program may have saved money and time in some cases, but in other cases it resulted in maintenance issues being exacerbated and even injury when an untrained person tried to fix an issue without the correct tools. Behavioral science can provide managers and executives with the knowledge and skills to better understand the antecedents, barriers, and consequences created by their verbal behavior and decisions. This can lead to a better identification of contributing factors and solutions.

Metacontingencies In process safety

In process safety, the link between leader behavior and the event is often latent and difficult to detect (Bell & Healey). Research has shown that in situations involving a series of decisions, each person may only assess their risk according to their action without understanding how it links to other decisions and increases risks for other entities (Watkins & Bazerman, 2003). This can be partially remedied by being able to predict the downstream impact of a leader's own behavior and decisions (i.e., metacontingencies; Glenn, 1988), both potentially increasing the saliency of the interlocking contingencies (Glenn & Malott, 2004) and potential risk. Doing so likely requires a clear understanding of how antecedent and consequences influence behavior as well as a system for obtaining regular upward feedback.

A significant amount of managers and executives' influence on the workplace occurs during meetings, informal conversations, and high-level decision making. Subtleties in their verbal behavior matter and tools for evaluating and improving behavior in those contexts are needed (Houmanfar, Rodrigues, & Smith, 2009). Behavioral science has much to offer for positively impacting leader verbal behavior related to safety. One study improved supervisor safety verbal behavior by giving them feedback from the manufacturing workforce (Zohar, 2002). This intervention provided the supervisors with the flexibility to decide on the safety-related content to discuss while encouraging more interactions in general. As supervisors increased the amount of their safety interactions with the workforce, the safety performance among their teams dramatically improved. However, the study only classified the conversations reported as related to safety, production, or both but did not measure other elements of the conversations nor did it seek to understand the impact through layers of management and on specific process safety issues.

Leaders make decisions on strategy, budget, and priorities, among other things, that create what can be referred to as metacontingencies (Glenn & Malott, 2004). Those metacontingencies, at least in part, put into place the environments that then produce the antecedents and consequences influencing worker behavior, called interlocking behavioral contingencies (Glenn & Malott, 2004). These interlocking behavioral contingencies can result in a work environment (e.g., immediate supervisor says, "we really need to pick up production, we're falling behind on customer orders"), physical environment (e.g., new equipment purchased making work easier and safer), task environment (e.g., a lockout/tagout procedure to de-energize equipment before working on it that includes many steps to perform safely and yet shortcuts are sufficient to achieve the goal), or surroundings (e.g., severe weather when scheduled to conduct work outside) that increases or decreases the likelihood of behaviors that support or degrade process safety controls. The impact of leader behavior cascades throughout the organization to the frontline works and impact not only their results, but whether they attained those results safely. Figure 1 demonstrates that each layer of management creates a work environment that influences the next layer until the environment has been created for the people directly delivering the product or results. This means an understanding of behavior and the work environment created by those behaviors is needed at each level of an organization in order to more comprehensively impact the safety of frontline employees.

Process safety is often thought of as an engineering problem with engineering solutions. However, for engineered controls to be effective, people throughout the organization still have to behave in critical ways. Managers and executives are in a position to reinforce these behaviors at a high rate. Managers and executives make decisions every day that influence the physical environment, task, surroundings, and the behaviors of people who report to them through their staffing, structure of the organization, team assignments, coaching, or lack thereof, duration content and attendance at meetings, quarterly goals, communication practices, strategic focus, bringing in new product lines, and so on. It makes sense that a process safety intervention must involve top level leaders to be successful.

It is often difficult for managers and executives to link their day-to-day actions and decisions to the safety performance of the frontline supervisors and workers and the current environment because (a) the consequences of

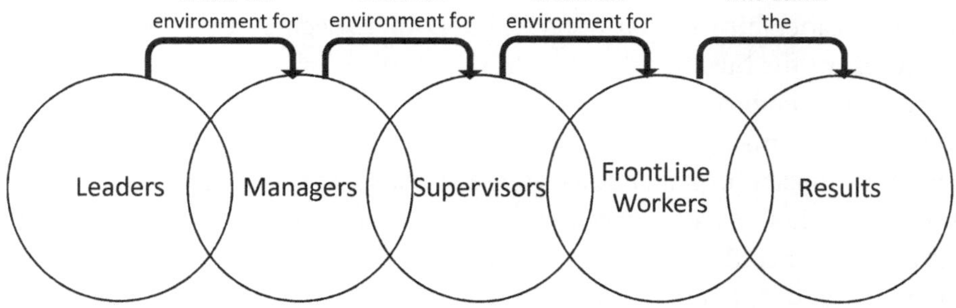

Figure 1. This figure depicts the cascading effect of the environment that leaders create in an organization referred to as the consequence chain.

their decisions may be too delayed (Smallman & John, 2001), (b) the interlocking chain of behaviors are too remote or diffused to be salient to the decision maker (Hopkins, 1999), (c) lack of leading measures and focus on lagging measures result in responding only after an event has occurred (National Safety Council, 2015), (d) and immediate consequences may distract leaders from safety (Culter & James, 1996). Most of the powerful, immediate consequences experienced by managers and executives come from their peers, the company board, production results, and major events (Hyten & Ludwig, 2017). This can produce mostly reactive or short-sighted decision making based on financial criteria instead of maintaining control over process safety issues such as the behavior of their people interacting with the fallout of these decisions. For example, if the board demands budget cuts and a site director delays updating dilapidated equipment, this may result in employees taking short cuts, ignoring signs of potential failure, and creating work arounds that lead to increased risk. The spending hold, while meant to reduce costs and please the board, may inadvertently communicate to employees that safety is not important to leadership and could result in equipment failures and increased risk. In the BP disaster, reports state that short cuts to save time and money for the over budget project were enacted by leaders and those decisions lead to a series of events that resulted in the explosion and spill (United States Government Publishing Office, 2011). Of course, this was not their intention and they did not foresee the impending consequences. Identifying the metacontingencies that link leader behavior to the environment created for frontline employees could prevent situations like this.

Organizational strategy

Perhaps the starting point for identifying metacontingencies and the subsequent interlocking behavioral contingencies that impact process safety is organizational strategy. All organizations have a purpose and reason for existing (Handy, 2002). If these companies do not run a viable and profitable business, they will not survive. A business strategy or philosophy is there to guide the leadership company stakeholders in a specific direction to support the success of the business. Leaders use strategy or "philosophical statements" to create the environment where workers perform and produce organizational results (Smith & Chase, 1990).

Heavy industry organizations whose workers are exposed to hazards or whose operations have the potential to cause harm to many are subject to a multitude of health and safety regulations to protect the company, the individual and the public from the exposure to hazards (OSHA, 1999, 2002). These hazards and regulations must be considered at the strategy level because major failures can put the organization's viability at risk. The job of leadership, whose company operating methods generate health and safety risks, is to set up the environment to enable people to deliver on the direction or strategy, while controlling and reducing the risk to employees, workers, and members of the public.

Some health and safety initiatives, including process safety, are selected and implemented at lower levels of leadership in an organization and therefore, are not integrated into business strategy. Process safety initiatives may indeed be in opposition to some business strategies and a trade-off between the two must occur (O'Dea & Flin, 2003). Research on process safety incidents suggests that pressure for productivity and performance results in management encourage practices that are in conflict with safety policies and procedures (Cutler & James, 1996). To manage this struggle between productivity and safety, organizations create a separate safety department, separate safety documentation, and separate safety procedures all attempting to override the consequences created by the financial strategy. Because process safety issues are so intimately related to strategies and decisions that impact quality and production, process safety should receive the same attention that quality and production do at the strategy level of an organization and at the same time. This would help align the consequences for the frontline workers with productivity *and* safety.

Creating the environment for a safe and productive workplace starts at the strategy level. Leaders could significantly improve process safety as well as quality and production if they created a strategy built and designed using a fundamental understanding of how work environments influence critical process safety behaviors. This strategy would specify the role that leaders play in the creation of an environment that supports the behaviors critical to process safety. In other words, metacontingencies become prominent in organizational strategy and are formulated based on their impact on process safety. This would result in a more realistic path with achievable goals and more accountability for high-level leaders to help create the environments that enable these goals to be achieved. Organizational leaders equipped with a basic understanding of behavioral science could outline the antecedents and consequences created by organizational objectives from the top of the organization through the frontline employees. When an objective is not being met, the leaders could then easily trace the issue up through the consequence chain (Figure 1) to determine where adjustments need to be made.

Measurement

To create, evaluate, and improve an effective strategy, good measures must be in place to understand what is currently happening in the organization and to determine if things are getting better or worse. Many organizations track outcome measures like injuries and production and some track measures that are slightly more leading (i.e., occurs before an event) like close calls/near misses. It is notable that most of the feedback and consequences managers and executives receive from these safety measures are based on failure. Incident investigations used to identify and suggest remedies for failures tend to only focus on the local environment where the failure occurred, which is probably due to the fact that investigations are often run by supervisors and managers who oversee that environment (OSHA, 2015). This makes it difficult to trace the failings back to the leadership behaviors that created the environment except in the case of a catastrophic event, when more intensive investigations are conducted by outside officials, which is too late. But, most important, if improvements or changes are only determined after something has gone wrong, they miss out on a myriad of opportunities to improve the work environment and prevent incidents, including catastrophic ones.

Measures that examine the conditions in the context of metacontingencies may produce solutions that impact a wide range of safety issues, including process safety ones, like management involvement, ability to upward report issues and concerns honestly, and availability of appropriate equipment (Ludwig, 2017). BBS provides a source of leading measures and information but it often does not make its way up through the interlocking contingencies in the organization. For example, BBS observations may identify that employees are not regularly wearing hearing protection. Peer and group feedback as well as making hearing protection more readily available may quickly remedy this specific concern. But, the concern that managers had probably regularly observed employees without hearing protection and had not effectively addressed this situation prior to BBS remains. The managers may not have the skills to encourage safety behaviors and they may not understand that high response effort discourages safety behaviors. This means that other safety concerns are undoubtedly lurking. Through better incident investigations and improved leading measures, leaders can use the information to self-manage their own behavior and become more effective at positively impacting safety over time.

Proactive, success-based measures that contribute to process safety may include measures of maintenance completion and equipment functioning, employee compliance with procedures, work request submissions, employee reporting of concerns, money spent efficiently and effectively on safety improvements, employee competency, management of change, and measures on responses to nonstandard events that arise. Even something as simple as feeding back safety communication and safety climate scores to managers may be enough to encourage improved communication and use of behavioral science techniques at every level of an organization (Cooper, 2006; Kines et al., 2010; Zohar & Luria, 2003; Zohar & Polachek, 2014). Measurement and information gathering techniques must support honest reporting and quick resolution, which can also be measured through work completion time and employee perception surveys. It is important to note that the same measurement strategy will not work for every organization because each is unique and therefore, managers must have a good enough understanding of how the work environment influences behavior to select measures that fit their unique organization.

A construction organization, Costain, was recently reaccredited by the Cambridge Center for excellence in behavioral safety Costain (Cambridge Center for Behavioral Studies, 2015). Over time the measures they use have evolved and become more sophisticated. They go beyond traditional safety and BBS observations and include obtaining feedback from the workforce and project sites demonstrating that they are using behavioral principles to improve safety through improvement projects, implementation plans, training progress, and engagement scores. They roll all of their data into a scorecard so that each site's safety-related activities can be monitored. Sites are also encouraged to create measures specific to their environment and performance deficits as part of improvement projects. Project topics cover a wide range including wearing fall protection, increasing use of designated walkways, and avoiding service strikes while digging. This level of customization encourages each site to be actively engaged at identifying their own opportunities for improvement and finding ways to change. The sophisticated measures provide indication about the overall management of safety on each site and when additional supports may be required. This is only possible through a fundamental understanding of behavioral science at the leadership level and a flexible measurement system.

An effective behavioral approach to process safety will require understanding that manager and executive behavior is an important factor and should be managed and measured as much as or more than employee behavior. Leaders create the work environment for managers who encourage and discourage certain behaviors and this goes down the chain. Measures should be developed to evaluate the organization at all levels. At the worker level, conducting safety observations, pointing out hazards, and providing sideways and upward feedback related to safety should be measured. At the supervisor level, data evaluating communication of safety data, hazards, and fixes, and feedback from direct reports on effectiveness and responsiveness should be collected. Managers and directors should be evaluated on how effectively their people are managing safety as well as identifying issues and implementing fixes. In addition, upward feedback on their effectiveness and responsiveness should be collected and delivered on a regular basis. Use of behavioral and engineering tools to continuously evaluate and improve safety could be monitored at every level of the organization. The ongoing upward feedback as well as data collection on hazards, close calls, and communication should allow leaders to detect potential safety issues and fix them before they become an issue. A BBS observation may not capture time pressure causing people to take short cuts but a climate survey could. Seeing all safety measures together on a scorecard could also draw attention to leaders who are meeting expectations and those who need more support and coaching when they are lagging behind on several measures.

Creating a measurement system to properly evaluate leaders' influence on process safety is quite an undertaking. This can only be accomplished if leaders learn enough behavioral science to be able to pinpoint behaviors, measure, and identify consequences and antecedents that encourage or discourage those behaviors interlocking through leaders and managers to supervisors and workers.

Teaching leaders behavioral science

Leaders are no different than others; they are influenced by the antecedents and consequences in their environment. But they also have more power to arrange their own environment and the environment of others to obtain the best performance and results. One reason a scientific approach is so powerful is that it can help leaders see the role of their behavior in influencing coworkers and subordinates. Behavioral science provides leaders with techniques and a model to modify their actions and observe the impact of these changes. Yet, very few leaders know much about the concepts and techniques of behavioral science or how to effectively observe and change their own behavior and the behavior of others.

Because high-level leader behavior is dynamic (i.e., their jobs involve a wider range of behaviors and evolve based on the needs of the business much more so than the average frontline worker) and the business landscape is continually shifting, measurement systems need to allow for flexibility. Teaching leaders about behavioral science can help them create a more dynamic process and tools for designing metacontingencies that shape behavior throughout the organization in ways that would enhance controls of process safety issues. For example, as equipment begins to reach its life cycle and increased maintenance is required, a manager who understands behavioral science can install antecedents and consequences that encourage submitting work requests, speaking up when work arounds start to occur, and helping maintenance employees complete these tasks safely, completely, and in a timely fashion. The leader could also add measures to the scorecards to track these activities. And, executives can craft a strategy and make budgeting and hiring decisions that support ensuring that employees are trained and competent to complete their jobs and respond appropriately in emergency situations.

Leaders do not need an advanced degree in behavioral science to be able to apply it in their span of influence. In fact, many leaders may be able to influence their own behavior and the behavior of others simply by learning to pinpoint, measure, change, and evaluate (Nordstrom, Lorenzi, & Hall, 1990;

Snyder, 2006). OBM has already developed an approach called the Consultant Workshop Model for teaching leaders about behavioral science and coaching them to apply it (Ackley & Bailey, 1995; Godat & Brigham, 1999; Nordstrom et al., 1990). This model teaches by coaching leaders through their own performance improvement projects and supporting them to change behavior. This approach has also been used to teach university students performance management skill (Ackley & Bailey, 1995), managers to self-manage work-related issues (Godat & Brigham, 1999), and city government managers (Nordstrom et al., 1990) and school administrators (Maher, 1984) to improve employee performance. Participants in these studies were taught basic OBM concepts like pinpointing and ABC Analysis and were also coached to conduct a behavior change project. A similar model was used by Costain (Cambridge Center for Behavioral Studies, 2015) to create the system mentioned earlier. This practice equips leaders with the skills to continue improving performance after the consultation period so they may be more readily able to adequately modify and sustain the performance management program according to organizational needs (e.g., Snyder, 2006). The research studies published on this approach are relatively old and more follow-up regarding whether this approach produces lasting impact is needed, yet it has promise for applying to process safety.

Several behavioral science concepts and applications are suggested for inclusion in the training and coaching. Based on the causes previously identified by Bell and Healey (2006), some behavioral principles and techniques that may be of value for leaders to learn are listed in Table 1. For example, insufficient or ineffective management practices could be improved by teaching leaders to pinpoint, self-management, assess performance using the Performance Diagnostic Checklist (PDC), and having them gather upward feedback on a regular basis. Many of those listed are repeated or overlapped and others could be added. Research could further refine this list and produce a learning hierarchy.

These concepts may be best taught in context with examples from both process safety and the organization where the leaders work to bolster application and generalization. For example, learning to pinpoint behaviors associated with situational awareness (e.g., checking, inspecting), measure them, and then improve them through antecedents, feedback, and consequences, could be enough to have a substantial impact on process safety. In addition, assessments like ABC Analysis (Daniels & Bailey, 2014) and PDC (Austin, 2000) can be used to help leaders better understand the environment and how it may encourage or discourage behaviors associated with process safety in any context, including meetings, toolbox talks, or on the shop floor. In other areas of applied behavior analysis, assessment is common practice before an intervention program is selected and OBM should be no different. Teaching leaders to understand the environmental variables influencing behavior could result in much more

Table 1. Behavioral Principles and Techniques That Leaders Could Learn to Directly Impact Causes of Process Safety Issues.

Cause Identified by Bell and Healey (2006)	Behavioral principles and techniques that may be useful for leaders to learn
Ineffective or insufficient management practices	• Pinpointing • Self-management • Upward feedback • Performance Diagnostic Checklist (PDC)
Poor systems for managing safety	• Process mapping and process improvement • Measuring behaviour
Communication issues between employees and leaders	• Creating effective antecedents • Delivering feedback effectively • Measuring the impact of a change
Setting difficult goals for production, creating pressure to perform	• Setting goals effectively • Measuring the impact of a change • Understanding interlocking behavioral contingencies
Long work hours	• Measuring performance effectively • Upward feedback • Managing performance
Changing equipment or procedures without adequately informing or training employees	• Creating effective antecedents • Instructional design • Coaching and feedback
Insufficient or ineffective systems for reporting concerns	• Process improvement • Consequences • ABC Analysis and PDC • Understanding interlocking behavioral contingencies
Problems with maintenance work and completion	• Managing performance • Process improvement • Feedback
Poor systems to train and coach employees on knowledge and skills	• BST • Instructional design • Coaching • Feedback
Failure to adjust and improve after an incident occurs	• ABC Analysis and PDC • Feedback • Consequences • Understanding interlocking behavioral contingencies
Workers not following correct procedures	• Coaching • Antecedents and consequences • Understanding interlocking behavioral contingencies • ABC Analysis and PDC • Feedback

effective solutions when hazards and procedure deviations are identified. It may be useful for leaders to understand behavioral principles like positive and negative reinforcement, punishment, and extinction and examples of their impact in process safety as well. For example, if reports of safety concerns are met with extinction or punishment, they will be less likely to occur in the future.

As an example of leaders using behavioral science basics to impact process safety, the authors of this paper worked with a client who had issues with employees shutting off alarms on a control panel without investigating the alarm further. A manager wanted to fix this situation while he was learning behavioral science. Through interviews and observation he found that many of the alarms were "nuisance alarms," meaning, they were sounding in error and did not require investigation. A low signal (legitimate alarm) to noise (nuisance alarm) ratio can increase the likelihood of missing the signal (Wolf, Horowitz, Van Wert, Kenner, Place, & Kibbi, 2007) and ignoring a significant event. The manager had asked the workers to report nuisance alarms so that they could be corrected several times to no avail. As he learned behavioral science he was able to pinpoint behavior (number of reports submitted to address nuisance alarms) and results (number of alarms) and measure. Then, he changed the antecedents by printing the reporting form (instead of requiring computer entry) and sitting next to the employees and helping them fill it out when nuisance alarms occurred. As the alarms were fixed, he shared the fixes as well as the reduction in alarms as feedback to the employees until they contacted the natural contingencies of having noticeably less nuisance alarms occurring. At that point, the employees became more likely to report alarms without prompting by the manager. Hundreds of less nuisance alarms occurred each week after just a few months of reporting and fixing them. This small project potentially prevented a large scale process safety failure. In every organization, numerous opportunities to change target behaviors that impact process safety exist, yet, many leaders may not know how to influence behavior change.

In addition, teaching leaders to understand the impact of the environment and consequences on employee behavior will allow them to consider those influences when making decisions. Creating a strategy that optimized behavioral systems for process safety requires helping leaders understand the concepts of metacontingencies and interlocking behavioral contingencies. In other words, leaders must learn to recognize the antecedents and consequences they put into place when they make decisions and the subsequent behaviors that may follow.

Leaders can be taught to identify metacontingencies and interlocking behavioral contingencies through practice scenarios and site specific issues. Every hazard, close call, and incident investigation will be an opportunity to practice the skills. In addition, common behaviors that are detrimental to process safety issues like discouraging feedback or setting unrealistic goals can be reviewed and tools to reduce these issues can be discussed. Leaders could conduct behavior improvement projects that could focus on issues that impact process and personal safety, which will not only allow them to practice the tools but may also directly impact safety immediately. Over time, their application of behavioral tools will become more sophisticated and they will be able to make changes that impact their selected measures, including feedback and climate scores.

Process safety issues are complex and they need a high-level solution. Engaging leaders who create the environments where these issues happen is imperative for reducing their occurrence. Focusing on a few behaviors won't suffice, we need to fundamentally change the way managers and executives set strategies, engage with employees, make decisions, and impact the work environment. Teaching them to use behavioral science can achieve this goal. But, more research is needed to determine the best approach for teaching and coaching these skills, including the specific concepts and skills that need to be included. For example, research could evaluate whether teaching leaders to use the PDC is sufficient to improve decision making and development of strategies to influence behavior. OBM is well-positioned to increase its impact on safety by further developing and refining the Consultant Workshop Model and applying it to process safety. It is our hope that this paper will encourage researchers to revisit the Consultant Workshop Approach or other approaches for teaching leaders about behavioral science so that it can be applied to leadership and improving important organizational issues, such as process safety.

References

Ackley, G. B. E., & Bailey, J. S. (1995). Teaching performance improvement using behavior analysis. *The Behavior Analyst, 18*, 73–81. doi:10.1007/BF03392693

Austin, J. (2000). Performance analysis and performance diagnostics. In J. Austin, & J. E. Carr (Eds.), *Handbook of applied behavior analysis* (pp. 321–349). Reno, NV: Context Press.

Behavioral Safety Technologies. (2011). *New findings on serious injuries and fatalities.* Ojai, CA: Leading with Safety.

Bell, J., & Healey, N. (2006). The causes of major hazard incidents and how to improve risk control and health and safety management: A review of the existing literature. *Health and Safety Laboratory.* Retrieved from http://www.hse.gov.uk/research/hsl_pdf/2006/hsl06117.pdf.

Bureau of Labor Statistics. (2014). National census of fatal occupational injuries (preliminary results). Retrieved from http://www.bls.gov/news.release/pdf/cfoi.pdf

Cambridge Center for Behavioral Studies. (2015). Costain Ltd. achieve behavioral safety reaccreditation. Retrieved from http://www.behavior.org/resource.php?id=871

Cooper, D. M. (2006). Exploratory analyses of the effects of managerial support and feedback consequences on behavioral safety maintenance. *Journal of Organizational Behavior Management, 26*(3), 1–41. doi:10.1300/J075v26n03_01

Cutler, T., & James, P. (1996). Does safety pay? A critical account of the Health and Safety Executive Document: 'The Costs of Accidents'. *Work Employment & Safety, 10*(4), 755–765. doi:10.1177/0950017096104008

Daniels, A. C., & Bailey, J. S. (2014). *Performance management: Changing behavior that drives organizational effectiveness.* Atlanta, GA: Performance Management Publications.

Glenn, S. S. (1988). Contingencies and metacontingencies: Toward a synthesis of behaviour analysis and cultural materialism. *The Behavior Analyst, 11*, 161–179. doi:10.1007/BF03392470

Glenn, S. S., & Malott, M. E. (2004). Complexity and selection: Impications for organizational change. *Behavior and Social Issues, 13*(2), 89–106. doi:10.5210/bsi.v13i2.378

Godat, L. M., & Brigham, T. A. (1999). The effect of a self-management training program on employees of a mid-sized organization. *Journal of Organizational Behavior Management, 19*(1), 65–83. doi:10.1300/J075v19n01_06

Handy, C. (2002). What's a business for? *Harvard Business Review, 80*(12) 1–8.

Hopkins, A. (1999). *Managing major hazards. The lessons of the moura mine disaster*. Sydney, Australia: Allen & Unwin.

Houmanfar, R. A., Rodrigues, N. J., & Smith, G. S. (2009). Role of communication networks in behavioral systems analysis. *Journal of Organizational Behavior Management, 29*, 257–275. doi:10.1080/01608060903092102

Hyten, C., & Ludwig, T. (2017). Complacency in process safety: A behavior analysis toward prevention strategies. Journal of Organizational Behavior Management, 37(3–4), 240–260.

Ludwig, T. (2017). Process safety behavioral systems: Behaviors interlock in complex process safety meta-contingencies. Journal of Organizational Behavior Management, 37(3–4), 224–239.

Kines, P., Anderson, L. P. S., Spangenberg, S., Mikkelsen, K., Dyreborg, J., & Zohar, D. (2010). Improving construction site safety through leader-based verbal safety communication. *Journal of Safety Research, 41*, 399–406. doi:10.1016/j.jsr.2010.06.005

Komaki, J. L., & Citera, M. (1990). Beyond effective supervision: Identifying key interactions between superior and subordinate. *The Leadership Quarterly, 1*(2), 91–105. doi:10.1016/1048-9843(90)90008-6

Komaki, J. L., Desselles, M. L., & Bowman, E. D. (1989). Definitely not a breeze: Extending an operant model of effective supervision to teams. *Journal of Applied Psychology, 74*(3), 522–529. doi:10.1037/0021-9010.74.3.522

Maher, C. A. (1984). Training educational administrators in organizational behavior management: Program description and evaluation. *Journal of Organizational Behavior Management, 6*(1), 79–97. doi:10.1300/J075v06n01_06

McSween, T. A. (2003). The value-based safety process: Improving your safety culture with a behavioral approach. (2nd ed.). New York, NY: John Wiley & Sons, Inc.

McSween, T., & Moran, D. J. (2017). Assessing and Preventing Serious Incidents with Behavioral Science: Enhancing Heinrich's Triangle for the 21st Century. *Journal of Organizational Behavior Management, 37*(3–4), 283–300.

National Safety Council. (2015). Practical guide to leading indicators: Metrics, case studies, & strategies. Retrieved from: http://www.nsc.org/CambpellInstituteandAwardDocuments/WP-PracticalGuidetoLI.pdf

Ness, R. (2015). Lessons learned from recent process safety incidents. *CEP Magazine, 15*, 23–29.

Nordstrom, R. R., Lorenzi, P., & Hall, R. V. (1990). A behavioral training program for managers in city government. *Journal of Organizational Behavior Management, 11*(2), 189–211. doi:10.1300/J075v11n02_11

O'Dea, A., & Flin, R. (2003) The role of managerial leadership in determininsg workplace safety outcomes. *Health and Safety Executive Research Report RR044*. Retrieved from http://www.hse.gov.uk/research/rrhtm/rro44.htm.

Occupational Health and Safety Administration. (1999). Statement of Charles N. Jeffries assistant secretary for occupational safety and heath Retrieved from https://www.osha.gov/pls/oshaweb/owadisp.show_document?p_id=98&p_table=TESTIMONIES

Occupational Health and Safety Administration. (2000, reprinted). Process safety management. Retrieved from https://www.osha.gov/Publications/osha3132.html

Occupational Health and Safety Administration. (2002). Employer responsibilities. Retrieved from https://www.osha.gov/as/opa/worker/employer-responsibility.html

Occupational Health and Safety Administration. (2015). Incident (accident) investigations: A guide for employers. Retrieved from https://www.osha.gov/dte/IncInvGuide4Empl_Dec2015.pdf

Smallman, C., & John, G. (2001). British directors perspectives on the impact of health and safety on corporate performance. *Safety Science, 38*, 227–239. doi:10.1016/S0925-7535(01)00003-0

Smith, J. M., & Chase, P. N. (1990). Using the vantage analysis chart to solve organization-wide problems. *Journal or Organizational Behavior Management, 11*(1), 127–148. doi:10.1300/J075v11n01_09

Smith, W. K., Lewis, M. W., & Tushman, M. L. (2016, May). “Both/and” leadership. *Harvard Business Review, 94*(5), 62–70.

Snyder, G. (2006). The path to sales power. PM E-Zine. Retrieved October 9, 2007, from http://www.pmezine.com/article_dtls.asp?NID=255

Tuncel, S., Lotlikar, H., Salem, S., & Daraiseh, N. (2006). Effectiveness of behavior based safety interventions to reduce accidents and injuries in workplaces: Critical appraisal and meta-analysis. *Theoretical Issues in Ergonomics Science, 7*(3), 191–209.

Watkins, M. D., & Bazerman, M. H. (April, 2003). Predictable surprises: The disasters you could have seen coming. Harvard Business Review. Retrieved from https://hbr.org/2003/04/predictable-surprises-the-disasters-you-should-have-seen-coming/ar/1

Wolfe, J. M., Horowitz, T. S., Van Wert, M. J., Kenner, N. M., Place, S. S., Kibbi, N. (2007). Low target prevalence is a stubborn source of errors in visual search tasks. *Journal of Experimental Psychology: General, 136*(4), (623–638).

Zohar, D. (2002). The effects of leadership dimensions, safety climate, and assigned priorities on minor injuries in work groups. *Journal of Organizational Behavior, 23*(1), 75–92. doi:10.1002/(ISSN)1099-1379

Zohar, D., & Luria, G. (2003). The use of supervisory practices as leverage to improve safety behaviour: A cross-level intervention model. *Journal of Safety Research, 34*, 567–577. doi:10.1016/j.jsr.2003.05.006

Zohar, D., & Polachek, T. (2014). Discourse-based intervention for modifying supervisory communication as leverage for safety climate and performance improvement: A randomized field study. *Journal of Applied Psychology, 99*, 113–124.

COMMENTARY: Integrating Behavioral Science with Process Safety Management

Joseph Dagen and Marcin Nazaruk

In high-hazard industries the consequences for major incidents can be ultimate, the environment can require extensive remediation, and the economic impact can be significant and far reaching for both company and community. Most readers will be familiar with many of the highest profile major accidents, such as Apollo 1, Three Mile Island, Bhopal, Space Shuttle Challenger, Chernobyl, Piper Alpha, Exxon Valdez, Space Shuttle Columbia, and the Deepwater Horizon, to name a few. These incidents often occur in industries of global economic importance, and therefore the significant impact of major accidents is felt on a global scale.

Countless professionals have dedicated their lives to learning from these incidents. Today, safety professionals work to predict and prevent major incidents by applying cutting-edge scientific, engineering, scientific, human factors (HFs), and leadership solutions to high-hazard industries. Through substantial interdisciplinary effort, many of these industries now operate with relatively low failure rates (IOGP, 2016).

However, process safety events large and small still continue to occur. Additional improvement in process safety performance requires continually improving safety theory, operational practices, and industry performance. Applied behavior analysis (ABA) has much to offer to the process safety community, and the papers in this special issue of the *Journal of Organizational Behavior Management* (*OBM*) are a welcomed entry into the conversation. Since the contribution behavioral science can make is substantial, the papers in this issue will be discussed as they intersect with modern incident causation theory and the discipline of HFs. Additional research suggestions are included for those seeking to build on the papers in this issue and contribute even more to reducing the likelihood and impact of major process safety incidents.

Since HFs emerged as a scientific discipline during World War II, safety practitioners have tried to optimize the interaction of people, machines, and processes (Meister, 1999). Rodriguez, Bell, Brown, and Carter's (2017) discussion of the integration of OBM and HFs demonstrates how behavioral science can impact process safety through existing channels. Successfully managing process safety risk requires tools, processes, and expertise in a variety of HFs topics. The list of HF topics by the United Kingdom Health Safety & Executive (H&S regulator), cited by Rodriguez et al. (2017), are one of many such lists. Another categorization published by the Center for Chemical Process Safety (CCPS), (Crowl, 2007), includes 28 topics with associated tools and methods considered important to process industries (see Table 1). Rodriguez et al. (2017) offered a number of tools that align to some of the 28 common HF topics. Behavior analysts interested in building on the work of Rodriguez et al. should consider every topic of the 28 as an opportunity to refine and enhance existing practices. This may mean entering areas of research where ABA has not been active to date, such as labeling or qualitative hazard analysis; but the challenge and outcome are worth the potential hurdles in jargon and understanding. For additional information about the current status of HFs in process industries, see Amyotte and Lupien (2017), Edmonds (2016), and Robb and Miller (2012).

Another clear example of how ABA can contribute meaningfully to the process safety community is the translation of the term "complacency" prepared by Hyten and Ludwig (2017). This term, broadly used by engineers, safety, and HFs professionals, is too vague in definition to help pinpoint critical behaviors. To reach the audience of process safety practitioners, the behavioral translation offered by Hyten and Ludwig should be further promoted in generic safety publications but without technical or scientific jargon. Otherwise, the benefit of the precise definition will only be available to the few professionals either trained in ABA or those receiving coaching from ABA professionals. Behavioral science, like all sciences, maintains strict definitions of terms, but this also can prevent a broader population from embracing the benefits ABA demonstrably brought to many businesses. Furthermore, we'd argue that there is a need to "market" ABA/OBM within

Table 1. A List of Human Factors Areas Identified by the Center for Chemical Process Safety.

Facilities and equipment chapters	Management systems chapters
Process safety equipment design	Safety culture
Process control systems	Behavior-based safety
Control center design	Project planning, design and execution
Remote operations	Procedures
Facilities and workstation design	Maintenance
Human computer interface	Safe work practices and permit to work systems
Safe havens—emergency response	Management of change
Labeling	Qualitative hazard analysis
People chapters	Quantitative risk assessment
Training	Safety systems
Communications	Competence management
Documentation design and use	Emergency preparedness and response
Environmental factors	Incident investigation
Workload and staffing levels	
Shift work issues	
Manual materials handling	

process safety engineering communities. This could be achieved by publishing easy to understand papers introducing the basics of behavioral science along with practical examples in such process safety journals as the *Journal of Loss Prevention in the Process Industries, Process Safety Progress*, and others.

The goal of process safety engineering is to uphold and enhance safe operations, which includes preventing catastrophic incidents. Therefore, an accurate understanding of how catastrophic incidents occur is fundamental to the introduction of comprehensive and sustainable control measures. Historically, incident causation theories suggested process safety failures were the products of causal chains of events (Attwood, Khan, & Veitch, 2006; Leveson, 2001), where event X causes Y, which causes Z. Modern incident causation theories, however, do not assume linear, causal relationships among events, but rather assume process safety failures emerge from interactions of elements of the work system (Dekker, 2004; Hollnagel, 2014; Leveson, 2004; Qureshi, 2008). These interactions change cumulatively over time to generate complex, unexpected, and unpredictable outcomes (EuroControl, 2014). The implications of this modern incident causation model are important for practitioners and researchers. For example, process safety failures cannot be solely attributed to the behavior of an operator or leader. Similarly, preventing failures and building improvement strategies must go beyond operator and leader behavior. Ludwig's example (2017) of applying behavioral systems analysis (BSA) to a process safety incident perfectly demonstrates this point.

The modern incident causation model also has implications for how behavioral science can improve process safety. Modern incident causation theory suggests high-hazard operations are variable and successful operations require continual adaption to constantly changing conditions; therefore, some degree of behavioral variability is necessary to successfully manage process safety risk (EuroControl, 2014; Hollnagel, 2012; Kontogiannis, 2010). Existing procedures may not be 100% correct and poor procedure designs, along with faulty procedure life-cycle management, are implicated in many instances of noncompliance. Because of this, adaption through variability allows for maintaining safety of the process (Hale & Borys, 2012a, 2012b). Lebbon & Sigurdsson's (2017) analysis of behavioral variability is a sound foundation to raise additional questions about both sources and the optimal degree of variability. Within a context of complex, large, dynamically changing organizations, studies of variability could seek to identify optimal levels across different teams and business functions.

Leadership itself is routinely identified across industries as a contributory variable in process safety failures (OEDC, 2012). Gravina, Cummins, and Austin's (2017) paper is a welcomed addition to the literature on leadership in high-hazard operations. Combined with teaching leaders to understand basic behavioral science, future behavior analytic research could adopt, integrate, and improve the literature on safety leadership and leadership in hazardous contexts (Kolditz, 2007). How results are achieved in high-hazard industries is as important as the results themselves (Alavosius, Herbst, Dagen, & Rafacz, 2014). Behavioral researchers could specify how leaders can inspire their employees to resist strongly reinforcing direct-acting contingencies (e.g., shortcuts, omitting steps in a procedure, accepting small deviations, etc.) and instead condition high-quality work as a potent reinforcer (Gravina, Cummins, & Austin, 2017). The discussion of ABA's contributions to the pay for performance discussion (Daniels, Daniels, & Abernathy, 2006) may also need to be revisited in the context of high-hazard operations.

The success of Gravina et al. at improving the response to alarms is commendable. ABA's impact on alarm management and more broadly on control room design and management could be tremendous by integrating the laws of behavior with industry laws and guides. To build on the success of Gravina et al., additional research should consider integrating behavioral science with alarm management as regulated by US law (49 CFR 195.446 - Control room management), and/or industry standard guides (ANSI, 2009; EEMUA, 2013). This would potentially enable hundreds of other businesses to benefit from ABA's contribution.

Finally, modern incident causation theory has implications for the application of the safety triangle as presented by McSween and Moran (2017). Despite criticism, Heinrich's triangle is still a very popular framework used across industries (Gallivan, Taxis, Franklin, & Barber, 2008; Manuele, 2002; Taxis, Gallivan, Barber, & Franklin, 2005). To bring McSween and Moran's improvements to process safety, future research should consider how McSween and Moran's insights can benefit existing process safety conversations. For example, this work could impact industry dialogue about leading indicators for catastrophic process safety failures. The CCPS report on process safety key performance indicators (CCPS, 2007) applied the safety triangle concept to Loss of Primary Containment (LOPC) events by placing the Tier 1 and Tier 2 incidents at the top and minor LOPCs along with other system failures (including unsafe behaviors) at the bottom of the triangle. We see a contributory overlap between these models for personal safety and process safety and additional work is merited.

Perhaps the greatest opportunity for behavior science to impact process safety is through partnerships of industry associations and regulatory agencies (e.g., IChemE, CCPS, Abnormal Situation Management Consortium, etc.). Many of these institutions and the companies they support could benefit from access to expertise in ABA. Behavioral scientists, through academic institutions or other recognized ABA agencies such as the Cambridge Center for Behavioral Studies (CCBS) or the Association of Behavior Analysis International (ABAI), could partner with these regulators and institutions to provide guidance for the use of behavioral science. The scope of influence could cover discrete recommendations to broad industry guidance. For example, the CCBS behavior-based safety (BBS) experts could help improve industry-recognized contingency analysis tools that contain technical inaccuracies such as claiming antecedents can be present but not effective. Because antecedents are functionally defined, identifying ineffective antecedents ultimately leads to analyses that are unlikely to allow practitioners to change behavior. At the macrolevel, the CCBS or ABAI could contribute meaningfully to discussions of process safety regulation by applying expertise in BSA. Ludwig's (2017) analysis of metacontingencies and interlocking behavioral contingencies could form a foundation for these regulatory contributions.

As articulated by Baer, Wolf, and Risley (1987),

> increasing our effectiveness will not be easy, and it will not happen quickly. We should expect a long period of difficult, expensive, repetitive, and sometimes ineffective research into these applications, and we should enter that research with our best social skills, because we shall require the cooperation of unusually many people, often in unusually exposed positions. However, even with relatively little reaction-to-failure work behind us, it seems clear that we can do it.

The articles in this special issue are a great step toward bringing applied behavioral science to the process safety community; and those with an understanding of ABA who work in high-hazard industries eagerly welcome ABA's systematic entry into this globally relevant domain. There is much work to do.

References

Alavosius, M., Herbst, S., Dagen, J., & Rafacz, S. (2014). "Beyond the skinner box": Expanding behavior systems analyses. *Journal of Organizational Behavior Management, 34*(February 2015), 255–264. doi:10.1080/01608061.2014.973633

Amyotte, P. R., & Lupien, C. S. (2017). Elements of process safety management. In *Methods in chemical process safety* (Vol. 1, pp. 87–148). Cambridge, MA: Elsevier. https://doi.org/10.1016/bs.mcps.2017.01.004

ANSI. (2009). *ANSI/ISA-18.2-2009 management of alarm systems for the process industries*. Research Triangle Park, NC: The International Society of Automation.

Attwood, D., Khan, F., & Veitch, B. (2006). Occupational accident models--Where have we been and where are we going? *Journal of Loss Prevention in the Process Industries, 19*(6), 664–682. doi:10.1016/j.jlp.2006.02.001

Baer, D. M., Wolf, M. M., & Risley, T. R. (1987). Some still-current dimensions of applied behavior analysis. *Journal of Applied Behavior Analysis, 20*(4), 313–327. doi:10.1901/jaba.1987.20-313

CCPS. (2007). *Process safety leading and lagging metrics*. Retrieved from https://www.aiche.org/sites/default/files/docs/embedded-pdf/CCPS_ProcessSafety2011_2-24-web.pdf

Crowl, D. A. (2007). *Human factors methods for improving performance in the process industries*. Hoboken, NJ: Wiley-Interscience.

Daniels, A. C., Daniels, J., & Abernathy, B. (2006). The leader's role in pay systems and organizational performance. *Compensation & Benefits Review*, *38*(3), 56–60. doi:10.1177/0886368706288217

Dekker, S. (2004). Why we need new accident models. *Human Factors and Aerospace Safety*, *4*(1), 1–18.

Edmonds, J. (2016). *Human factors in the chemical and process industries: Making it work in practice*. Oxford, UK: Elsevier Science.

EEMUA. (2013). *Alarm systems--A guide to design, management and procurement*. London: Engineering Equipment and Materials Users' Association.

EuroControl. (2014). Systems thinking for safety: Ten principles a white paper. *moving towards safety-II*. Retrieved from http://skybrary.aero/bookshelf/books/2882.pdf

Gallivan, S., Taxis, K., Franklin, B., & Barber, N. (2008). Is the principle of a stable heinrich ratio a myth? *Drug Safety*, *31*(8), 637–642. doi:10.2165/00002018-200831080-00001

Gravina, N., Cummins, B., & Austin, J. (2017). Leadership's role in process safety: An understanding of Behavioral science among managers and executives is needed. *Journal of Organizational Behavior Management*, *37*(3–4), 316–331.

Hale, A., & Borys, D. (2012a). Working to rule, or working safely? Part 1: A state of the art review. *Safety Science*, 1–15. doi:10.1016/j.ssci.2012.05.011

Hale, A., & Borys, D. (2012b). Working to rule or working safely? Part 2: The management of safety rules and procedures. *Safety Science*. doi:10.1016/j.ssci.2012.05.013

Hollnagel, E. (2012). *FRAM: The functional resonance analysis method: modelling complex socio-technical systems*. Farnham, England: Ashgate Publishing Limited.

Hollnagel, E. (2014). *Safety-I and Safety-II: The past and future of safety management*. Burlington, VT: Ashgate Publishing Company.

Hyten, C., & Ludwig, T. (2017). Complacency in process safety: A Behavior analysis toward prevention strategies. *Journal of Organizational Behavior Management*, 37,(3–4), 240–260.

IOGP. (2016). *Safety performance indicators. Process safety events--2015 data*. London: International Association of Oil and Gas Producers.

Kolditz, T. A. (2007). *In extremis leadership: Leading as if your life depended on it*. San Francisco, CA: Jossey-Bass.

Kontogiannis, T. (2010). A contemporary view of organizational safety: Variability and interactions of organizaitonal processes. *Cognition, Technology & Work*, *12*(4), 231–249. doi:10.1007/s10111-009-0131-x

Lebbon, A., & Sigurdsson, S. (2017). Behavioral perspectives on variability in human behavior as part of process safety. *Journal of Organizational Behavior Management*, *37*(3–4), 261–282.

Leveson, N. (2001). *Evaluating accident models using recent aerospace accidents, part 1: Event-based models*. Retrieved from http://sunnyday.mit.edu/accidents/nasa-report.pdf

Leveson, N. (2004). A new accident model for engineering safer systems. *Safety Science*, *42*(4), 237–270. doi:10.1016/S0925-7535(03)00047-X

Ludwig, T. (2017). Process safety behavioral systems: Behaviors interlock in complex process safety meta-contingencies. *Journal of Organizational Behavior Management*, *37*(3–4), 224–239.

Manuele, F. A. (2002). Heinrich revisited: Truisms or myths. National Safety Council. Retrieved from http://books.google.co.uk/books?id=N-0OAAAACAAJ

Meister, D. (1999). *The history of human factors and ergonomics*. Mahwah, NJ: Taylor & Francis.

McSween, T., & Moran, D. (2017). Assessing and preventing serious Incidents with behavioral science. *Journal of Organizational Behavior Management*, *37*(3–4), 283–300.

OEDC. (2012). *Corporate governance for process safety* (p. 23). Paris: Organisation for Economic Co-operation and Development.

Qureshi, Z. H. (2008). *A review of accident modelling approaches for complex socio-technical systems* (No. DSTO-TR-2094). Edinburgh. Retrieved from http://crpit.com/confpapers/CRPITV86Qureshi.pdf

Robb, M., & Miller, G. (2012). Human factors engineering in oil and gas--A review of industry guidance. *Work, 41*(Supplement 1), 752–762. doi:10.3233/WOR-2012-0236-752

Rodriguez, M., Bell, J., Brown, M., & Carter, D. (2017). Integrating behavioral science with human factors to address process safety. *Journal of Organizational Behavior Management, 37*(3–4), 301–315.

Taxis, K., Gallivan, S., Barber, N., & Franklin, B. D. (2005). *Can the Heinrich ratio be used to predict harm from medication errors? Report to the Patient Safety Research Programme (Policy Research Programme of the Department of Health)*. Retrieved from http://eprints.pharmacy.ac.uk/764/1/BarberMedication_Errors.pdf

COMMENTARY: Is Organizational Behavior Management Enough? How Language and Person-States Could Make a Difference

E. Scott Geller

This unique collection of academic presentations of applied behavioral science addresses a real-world issue of critical importance--the effective maintenance of process safety to prevent disasters like NASA's Space Shuttle Challenger explosion in 1986 and the Deepwater Horizon oil spill in 2010. Bogard, Ludwig, Staats, and Kretchmen (2015) introduced the topic of process safety to readers of the *Journal of Organizational Behavior Management (JOBM)* with a "call to understand the contingencies involved in process safety" (p. 7), especially those contingencies contributing to "normalization of deviance" (Vaughan, 1996).

Each of the articles in this collection answer this "call" quite effectively, essentially by specifying various three-term contingencies that jeopardize process safety and suggesting strategies for modifying or adding contingencies to enhance process safety. Ludwig (2017a), Hyten and Ludwig (2017), and Lebbon and Sigurdsson (2017) explain "normalization of deviance" or the acquisition of complacency with reference to short-cutting behavior, organizational metacontingencies, habituation, and the avoidance paradox. McSween and Moran (2017) and Gravina, Cummins, and Austin (2017) explicate how the procedural steps of behavior-based safety (BBS) can be expanded and adapted to address safety-related behavior at all corporate levels, from the front line to the board room. In addition, Rodriguez, Bell, Brown, and Carter (2017) demonstrate how the practice of human factors can benefit from behavioral systems analysis, fluency-based training and development, and a behavioral analytical approach to leadership decision making and corporate culture. These authors claim that BBS and fluency-based training provide "behaviorally-sound actionable operations" to such human dispositions as competence, attention, personality, attitude, job enrichment, and risk perception.

A select audience

Each of the presentations voice a consistent concern that employers at all levels of an organization understand the behavioral challenges of process safety and implement behavior-focused tactics to address these issues, from redesigning behavioral checklists and promoting appropriate checklist use to educating company leadership about contingency management and motivating its application to decision making regarding process safety. Question: Does any reader believe the average corporate manager will be able to develop and apply a behaviorally-sound action plan from these academic presentations? Okay, so this is an unfair question because this scholarship was prepared for a professional research journal, not for direct real-world application. But where does this critical dialogue go from here?

Will readers be able to translate the technical jargon of these thoughtful presentations into meaningful language for a practitioner who has not received advanced education in organizational behavior management (OBM)? For example, read again the first sentence of Tim Ludwig's conclusion about the interlocked behavioral contingencies within a complex behavior-focused system that must be understood and applied toward the development of a practical action plan for process safety (2017b). Frankly, I do not believe my attempt at such a translation for a concerned CEO or safety professional would do these words justice. Yet these practitioners are the very individuals who need to understand and use the principles and recommendations detailed in this scholarship on process safety. It is worth noting that each of these presentations are authored by OBM professionals who consult with organizations regularly regarding practical applications of behavioral science technology.

Inconsistent language

Beyond the complex academic jargon, these presentations reveal inconsistencies in OBM language, which can stifle the appreciation and effective application of recommended procedures. In particular, I was most disappointed to see the term "accident" used by Lebbon and Sigurdsson (2017) and Rodriguez, Bell, Brown, and Carter (2017). The term "accident" implies the occurrence of a chance event with limited behavioral control. Thus, for more than a decade safety professionals have intentionally avoided using this expression and often substitute the word "incident," as the authors of the other presentations have done. However, I believe the term "incident" can sound too minor or insignificant, as do the related words: "incidental" and "incidentally." Hence, for more than four decades I have advocated using words that reflect the seriousness of the mishap (e.g., property damage, injury, fatality).

Also discouraging is the seemingly random interchange of the terms "behavior analysis" (Rodriguez et al., 2017) and "behavioral science" (Gravina et al., 2017). While "behavior analysis" reflects the historical roots of OBM, "behavioral science" implies more than an "analysis" of behavior and is a term more highly regarded by the public. Furthermore, recent textbooks advocate for the academic label "applied behavioral science" (Biglan, 2015; Geller, 2016a).

Also, it is disconcerting and confusing to behold the authors' varied and inconsistent reference to "positive reinforcement," a most popular term in both behavior analysis and behavioral science. Technically, positive reinforcement is the application of a pleasant consequence to increase the form or frequency of the behavior it follows. If the preceding behavior does not improve, the procedure was not "positive reinforcement" and the consequence was not a "positive reinforcer." Some of these presentations deviate from the professional definition of "positive reinforcement," and seemingly use the colloquial meaning of "reinforcement"--a supportive consequence.

Hyten and Ludwig (2017) indicate a BBS system is designed to provide "positive reinforcement to frontline workers for observed safe acts" (p. 543); and McSween and Moran (2017) refer to BBS as an "assignment-observation-feedback-reinforcement process" (p. 366). It seems these authors consider supportive behavioral feedback to be a positive reinforcer. However, it is possible supportive feedback has no direct impact on behavior, especially when delivered within the context of corrective feedback. Analogously, Lebbon and Sigurdsson (2017) advise practitioners to implement "praise and reinforcement" for sanctioned process safety responses to offset natural reinforcement received from the nonsanctioned responses (p. 685). Unfortunately, the authors provide no examples for such contrived "positive reinforcement."

Finally, the labels "leader" and "manager" are used interchangeably (Gravina et al., 2017; McSween & Moran, 2017), yet industrial/organizational psychologists and OBM researchers have specified operational distinctions between these organizational terms. For example, managers (or "transactional leaders") typically apply an extrinsic accountability system "to keep people on track to reach existing group or organizational goals" (Foti & Boyd, 2016, p. 273), whereas leaders (considered to be "transformational") "motivate followers beyond immediate self-interests ... while implementing principles of self-motivation" Foti & Boyd, 2016, p.275.

Daniels and Daniels (2014) explain why the measure of a leader is indicated by the amount of discretionary (or self-motivated) behavior s/he inspires the followers to perform; and Geller and Geller (2017) review the research relevant to increasing people's self-motivation for injury prevention. A discussion of techniques to enhance an individual's self-motivation is beyond the scope of this commentary, but the fact that behavioral science has defined interpersonal and environmental conditions that enhance self-motivated or self-directed discretionary behavior (Watson & Tharp, 1997) is good news for process safety.

Each presentation expresses the challenge of maintaining safety-related behavior when supportive extrinsic consequences are not evident. For example, Hyten and Ludwig (2017) end their presentation with reports of elite pilots sharing their feedback after completing their air shows, and of the most experienced space shuttle commander in NASA history keeping a log of personal errors per each mission. What contingencies motivated these professionals to perform these tedious and time-consuming behaviors? Where were the positive reinforcers? Why didn't these professionals develop complacency or succumb to the "avoidance paradox"?

Without an extrinsic accountability system, the pilots and astronaut were presumably self-determined to satisfy one or more of three basic psychological needs: (a) autonomy, (b) relatedness, (c) and/or competence (Deci, 1975; Deci & Flaste, 1995; Deci & Ryan, 1995; Ryan & Deci, 2000). Consider for example, the human need to be competent at performing worthwhile work, and how feedback can help to fulfill that need. Might this be the motive for the "Blue Angels" and Astronaut Jim Wetherbee systematically reviewing their professional behavior? Could the enhancement of perceived competence be the self-determination rationale for providing and receiving BBS feedback? Could the personal label of "competent safety leader" influence the occurrence of safety-process behaviors in the absence of a three-term contingency or an extrinsic accountability system? Accepting these possibilities requires an appreciation of a dispositional person-state and its impact on behavior.

The establishing operation of a person-state

Hyten and Ludwig (2017) emphasize that "complacency" is not a person-state that can account for human error, but rather "a descriptor for a changing pattern of behavior" (p. 55). Similarly, Crowell and Anderson (2004) criticized my use of certain person-states to define an individual's propensity to go beyond the "call of duty" on behalf of the safety and/or health of others (Geller, 1996, 2001, 2003). Crowell and Anderson refer to such person-states as mentalistic explanations or "explanatory fictions" (Baum, 1994; Michael, 1993) that can distract from a focus on improving overt behavior.

I hope readers will consider the potential value of a "person-state," if only as a label to define a moderator or an intervening variable that can function like an establishing operation (Michael, 1982) or an internalized rule (Malott, 1992) to activate the occurrence of certain behavior. Just as the physical states of food satiation and deprivation influence the impact of certain response-consequence contingencies, psychological dispositions or person-states can affect the relevance and influence of particular response-consequence contingencies. And, when certain environmental conditions or operations enhance particular dispositional person-states, the probability of relevant behavior occurring is increased. For example, situational factors that inspire a person-state of self-determinism (Deci & Ryan, 1995), self-persuasion (Aronson, 1999), or self-motivation (Geller, 2016b) increase the occurrence of self-directed behavior (Watson & Tharp, 1997)--discretionary behavior that occurs without an extrinsic three-term contingency. Since many safety-process behaviors need to occur in the absence of an extrinsic accountability system or reinforcement contingency, self-directed behavior is frequently required to support process safety.

As indicated earlier, empirical research has defined operational conditions that enhance the person-state of self-determinism or self-persuasion. Similarly, in discussions with employees, I've identified events, situations, and contingencies that enhance or stifle specific person-states that influence the occurrence of safety-related behavior (Geller, 1996, 2001; Geller & Geller, 2017). Granted, these discussions treat person-states as hypothetical constructs with substantial surplus meaning (MacCorquedale & Meehl, 1948), but these conversations define "behaviorally-sound actionable operations" (Rodriguez et al., 2017) that can be changed or implemented to increase the occurrence of safety-related behavior in the absence of relevant extrinsic contingencies.

Note that this discussion considered behavior to be an outcome of various situational and dispositional factors, and that is the perspective my partners at Safety Performance Systems, Inc. (safetyperformance.com) and I have voiced since 1994 when teaching BBS. Yet, McSween and Moran (2017) assert that "behavior is a function of the system, but it is not an outcome" (p. 251). This is likely only a misunderstanding of differential usage of the terms "process" and "outcome," but it does support a prominent theme of this commentary--the use of consistent user-friendly language to convey the principles and applications of behavioral science to those practitioners responsible for process-safety intervention and accountability.

Conclusion

A decade ago, Paul Chance (2007) reminded readers of *The Behavior Analyst* that B.F. Skinner's initial optimism about the science of behavior had shifted to pessimism, as evidenced by his 1982 address to the American Psychological Association: "Why we are not acting to save the world" (Skinner, 1987). Chance (2007) summarized various findings from behavioral science with particular reference to two familiar conclusions: (a) "Immediate consequences outweigh delayed consequences" (p. 154), and (b) "Consequences for the individual usually outweigh consequences for others" (p. 155). Chance concludes his treatise on the principles of human nature that suggest a pessimistic perspective regarding the human species with: "The ultimate challenge for behavior analysts is to prove B. F. Skinner wrong" (p. 153).

Each of these thoughtful perspectives on applying behavioral science to benefit process safety are true to the behavior analysis principles reviewed by Chance (2007) and therefore echo his conclusions. However, I propose adding the following four principles from psychological science that can be readily incorporated within the domain of OBM to advance process safety:

(a) person-states like self-persuasion, self-direction, self-accountability, and self-motivation are influenced by changeable situational factors and can activate the occurrence of behavior without the direct control of extrinsic reinforcement contingencies;
(b) Maslow's Hierarchy of Needs reflects person-states that connect to specific rewarding consequences; and consequences that satisfy higher needs or person-states (e.g., social acceptance, competence, and self-actualization) support the evidence-based determinants of self-motivation (Geller, 2015, 2017);
(c) the highest level of Maslow's revised need hierarchy is "self-transcendence" (Maslow, 1971)--reaching beyond individualistic self-needs to help others, which contributes to satisfying the needs (or person-states) for social acceptance, self-esteem, and self-actualization; and
(d) helping others, as when performing a safety-process behavior can be reinforcing because it contributes to satisfying the highest level of Maslow's Hierarchy of Needs and supports related person-states.

I expect pushback from OBM professionals regarding these four additional principles because they reflect aspects of psychological science that have been traditionally ignored by behavior analysts. But before disregarding this perspective on the establishing operations of the person-states reviewed here, please consider the potential process-safety behaviors that might be activated by a person-state like self-motivation, as well as the various situational and interpersonal factors that can enhance this dispositional person-state. If this commentary activates related conversation, more commentary, and relevant behavioral research, my person-state of "competence" will be flattered and in turn fuel my self-motivation to promote these perspectives further. Of course, a contrary reaction might occur and thus stifle my subsequent writing about person-states, at least for the *JOBM*.

References

Aronson, E. (1999). The power of self-persuasion. *American Psychologist, 54*, 875–884. doi:10.1037/h0088188

Baum, W. M. (1994). *Understanding behaviorism: Science, behavior and culture*. New York, NY: Harper Collins College Publishers.

Biglan, A. (2015). *The nurture effect: How the science of human behavior can improve our lives and our world.* Oakland, CA: New Harbinger Publications, Inc.

Bogard, K., Ludwig, T. D., Staats, C., & Kretchmen, D. (2015). An industry's call to understand the contingencies involved in process safety: Normalization of deviance and interlocking contingencies. *Journal of Organizational Behavior Management, 35,* 70–80. doi:10.1080/01608061.2015.1031429

Chance, P. (2007). The ultimate challenge: Prove B. F. Skinner wrong. *The Behavior Analyst, 30*(2), 153–160. doi:10.1007/BF03392152

Crowell, C.R., & Anderson, D.C. (2004). On reinventing OBM: Comments regarding Geller's proposals for change. Journal of Organizational Behavior Management, *24*(1/2), 27–53.

Daniels, A. C., & Daniels, J. E. (2014). *Measure of a leader: The legendary leadership formula for producing exceptional performers and outstanding results.* New York, NY: McGraw-Hill.

Deci, E. L. (1975). *Intrinsic motivation.* New York, NY: Plenum.

Deci, E. L., & Flaste, R. (1995). *Why we do what we do: Understanding self-motivation.* New York, NY: Penguin Books.

Deci, E. L., & Ryan, R. M. (1995). *Intrinsic motivation and self-determinism in human behavior.* New York, NY: Plenum.

Foti, R. J., & Boyd, K. B. (2016). Leadership, followership, and AC4P. In E. S. Geller (Ed.), *Applied psychology: Actively caring for people* (pp. 273–300). New York, NY: Cambridge University Press.

Geller, E. S. (1996). *The psychology of safety: How to improve behaviors and attitudes on the job.* Radnor, PA: Chilton Book Company.

Geller, E. S. (2001). *The psychology of safety handbook.* Boca Raton, FL: CRC Press.

Geller, E. S. (2003). Should organizational behavior management expand its content? *Journal of Organizational Behavior Management, 22*(2), 13–30. doi:10.1300/J075v22n02_03

Geller, E. S. (2015). Seven life lessons from humanistic behaviorism: How to bring the best out of yourself and others. *Journal of Organizational Behavior Management, 35*(1), 151–170. doi:10.1080/01608061.2015.1031427

Geller, E. S. (Ed.). (2016a). Applied psychology: Actively caring for people. New York, NY: Cambridge University Press.

Geller, E. S. (2016b). The psychology of self-motivation. In E. S. Geller (Ed.), *Applied psychology: Actively caring for people* (pp. 83–118). New York, NY: Cambridge University Press.

Geller, E. S. (2017). *Actively caring for people in schools: How to cultivate a culture of compassion.* New York, NY: Morgan James Publishers.

Geller, E. S., & Geller, K. S. (2017). *Actively caring for people's safety: Cultivating a brother's/sister's keeper culture.* Park Ridge, IL: The American Society of Safety Engineers.

Gravina, N., Cummins, B., & Austin, J. (2017). Leadership's role in process safety: An understanding of behavioral science is needed. *Journal of Organizational Behavior Management, 37* (3-4), 316–331.

Hyten, C., & Ludwig, T. D. (2017). Complacency in process safety: A behavior analysis toward prevention strategies. *Journal of Organizational Behavior Management, 37*(3-4), 240–260.

Lebbon, A., & Sigurdsson, S. (2017). Behavioral perspectives on variability in human behavior as part of process safety. *Journal of Organizational Behavior Management, 37*(3-4), 261-282.

Ludwig, T. D. (2017a). Process safety: Another opportunity to translate behavior analysis into evidence-based practices of grave societal value. *Journal of Organizational Behavior Management, 37*(3-4), 221–223.

Ludwig, T. D. (2017b). Process safety behavioral systems: Behaviors interlock in complex process safety meta-contingencies. *Journal of Organizational Behavior Management, 37*(3-4), 224–239.

MacCorquedale, K., & Meehl, P. E. (1948). On a distinction between hypothetical constructs and intervening variables. *Psychological Review, 55*, 95–107. doi:10.1037/h0056029

Malott, R. W. (1992). A theory of rule-governed behavior and organizational behavior management. *Journal of Organizational Behavior Management, 12*(2), 45–65. doi:10.1300/J075v12n02_03

Maslow, A. H. (1971). *The farther reaches of human nature*. New York, NY: Viking.

McSween, T., & Moran, D. J. (2017). Assessing and Preventing Serious Incidents with Behavioral Science: Enhancing Heinrich's Triangle for the 21st Century. *Journal of Organizational Behavior Management, 37*(3-4), 283–300.

Michael, J. (1982). Distinguishing between discriminative and motivational functions of stimuli. *Journal of the Experimental Analysis of Behavior, 37*, 149–155. doi:10.1901/jeab.1982.37-149

Michael, J. (1993). *Concepts and principles of applied behavior analysis*. Kalamazoo, MI: Society for the Advancement of Behavior Analysis.

Rodriguez, M., Bell, J., Brown, M., & Carter, D. (2017). Integrating behavioral science with human factors to address process safety. *Journal of Organizational Behavior Management, 37*(3-4), 301–315.

Ryan, R. M., & Deci, E. L. (2000). Self-determinism theory and the foundation of intrinsic motivation, social development, and well-being. *American Psychologist, 55*, 68–75. doi:10.1037/0003-066X.55.1.68

Skinner, B.F. (1987). Whatever happened to psychology as the science of behavior? American Psychologist, *42*(8), 780–786.

Vaughan, D. (1996). *The Challenger launch decision: Risky technology, culture, and deviance at NASA*. Chicago, IL: University of Chicago Press.

Watson, D. C., & Tharp, R. G. (1997). *Self-directed behavior: Self-modification for personal adjustment* (7th ed.). Pacific Grove, CA: Brooks/Cole Publishing.

COMMENTARY: Process Safety: Look Looking Beyond Personal Safety to Address Occupational Hazards and Risks

Oliver Wirth

> "A controlling agency, together with the individuals who are controlled by it, comprises a *social system* ... and our task is to account for the behavior of all participants."
>
> —B. F. Skinner in *Science and Human Behavior*, 1953, p. 335

Industrial disasters have the potential for grave consequences, not only for the personal well-being of the workers involved, but also for the well-being of the environment and its other inhabitants. Industrial organizations, despite their reliance on advanced engineering feats and highly mechanized or automated operations, are social systems comprised of individuals behaving in service to both organizational and personal interests. I believe the emergence of process safety management as a method for preventing industrial disasters is in part a realization that safety is a behavioral phenomenon. One only needs to peruse the investigation reports from recent high-profile industrial disasters (e.g., Chernobyl, Texas City and Deepwater Horizon) to see that combinations of human error, unsafe actions, poor decision making, and poor safety cultures are invariably cited as major causes (CSB, 2007, 2016; IAEA, 1992). Also evident from these reports is an apparent refocusing of the safety lens from the actions and events most proximal to the incident to include any and all upstream factors in the causal chain of events, such as the actions and decisions of contractors, supervisors, managers, executives, and any other individuals with organizational leadership responsibilities. It is perhaps no coincidence then that the term "accident" as a generic label for occupational safety failures has fallen out of favor among safety experts precisely because such failures are now seen to be wholly preventable (Mathis, 2013).

Therefore, I applaud the editors of the *Journal of Organizational Behavior Management* (*JOBM*) and its contributors for offering this special issue on *process safety* and encouraging researchers and professionals in the field of behavior analysis to answer the petrochemical industry's call for behavioral solutions to their process safety challenges (see Ludwig, 2017-a). The collection of articles survey and illustrate the variety of ways in which concepts, principles, and methods of behavior analysis can contribute toward the establishment and implementation of effective process safety programs. In hearing the industry's call for process safety solutions, I actually discern three desires: (a) clarification of concepts, (b) recommendations for practical and effective interventions, and (c) additional research.

A call for clarification

The U.S. Occupational Safety and Health Administration promulgated the standard titled Process Safety Management of Highly Hazardous Chemicals (PSM) in 1990 to prevent the catastrophic release of hazardous chemicals (OSHA, 2000). The standard emphasized the management of hazards and established a comprehensive management program that integrated technologies, procedures, and management practices. PSM established programmatic requirements across 14 different elements, such as hazard analysis, training, prestartup safety reviews, compliance audits, incident investigations, and emergency planning, to name just a few. Characteristic of many governmental regulations, the PSM standard specifies *what* elements or outcomes are needed in a PSM program, but not necessarily *how* to achieve them. Although various guideline documents exist to aid employers and safety professionals in implementing process safety management systems (e.g., AIChE, 2016; OSHA, 1994), these guidelines offer few specific or actionable solutions for addressing process safety challenges, particularly challenges associated with identifying, establishing, and maintaining the necessary behavioral supports necessary to achieve the desired PSM outcomes—enter behavior analysis.

Illustrating some of the basic behavioral processes that account for process safety failures is one way to establish the relevancy of behavior analysis to process safety. Toward this end, Hyten and Ludwig (2017) and Lebbon and Sigurdsson (2017) show how the basic concepts of habituation, positive reinforcement, negative reinforcement and avoidance, rule governance, stimulus control, and other higher-order behavioral processes can account for many behaviors that are either supportive or detrimental to process safety. The design of effective behavioral change interventions requires a thorough understanding of behavioral principles; however, these basic topics are not covered in most safety-related training manuals or curricula for safety professionals. Accordingly, Gravina, Cummins, and Austin (2017) recommended developing and testing a comprehensive safety leadership training curriculum that includes teaching the fundamentals of behavioral science.

Behavioral science can also help to identify the relevant causal variables contributing to safety failures. McSween and Moran (2017) provided a useful reconceptualization of the often-cited Heinrich's (1931) safety triangle, which describes the relationships among major injuries and fatalities, minor injuries, and near misses. McSween and Moran echo some of the criticisms of the triangle (e.g., Manuele, 2014), such as the lack of empirical evidence and predictive utility, but they also provide a potentially useful expansion and reconceptualization of the triangle to aid safety professionals in pinpointing at-risk behavior and its precursors, which include the leadership and system failures at different levels of an organization.

Another concept that still causes confusion is safety culture (and the related safety climate). In the wake of high profile industrial disasters, many organizations in high risk industries have placed a greater priority on conducting assessments of safety culture and implementing programs intended to create or foster a positive safety culture. Indeed, establishing a positive safety culture is considered a prerequisite for an effective PSM program (Kumar, 2014). Despite an appreciable increase in the number of published articles on safety culture in the last 25 years (Glendon, 2008), and arguably some progress in establishing its utility (Hofman, 2017; Schneider, 2017), safety culture is still not well defined nor well understood by industry leaders and safety professionals.

By illustrating the complex behavioral contingencies that can account for safety-related performances and decisions throughout an organization, Ludwig (2017-b) and Gravina et al. (2017) exemplify how behavior analysis can shed some much needed light on current conceptualizations of safety culture. Indeed, the behavioral analyses of cultures has received thoughtful analysis since B. F. Skinner initially provided a unified scientific theory of selection by consequences in which he illustrated the role of behavioral contingencies (e.g., reinforcement) in developing and maintaining cultural practices (Skinner, 1981). Behavioral interpretations of the origins and maintenance of cultures, and, notably, introduction of concepts such as metacontingencies and interlocking behavioral contingencies (Glenn, 1988; Glenn & Malott, 2004), provide a coherent framework that extends a behavioral analysis beyond the individual to the collective actions of social groups, societies, and cultures. Ludwig (2017-b) and Gravina et al. (2017) extend this framework convincingly to occupational safety, illustrating its descriptive power in identifying the behavior-based complex interpersonal, social, and cultural dynamics pertaining to occupational safety. Unfortunately, these concepts and framework have not yet permeated the mainstream safety culture literature. Linking behavioral accounts of cultural practices with process safety, which inherently goes beyond personal safety, is a good opportunity for academics, researchers, and professionals in the behavioral sciences to make important contributions to how safety culture is conceptualized, assessed, and influenced.

A call for effective solutions

Ongoing debates over the proper definitions of concepts, such as safety culture notwithstanding, safety professionals and industry decision makers mainly seek practical and effective solutions, and the articles in this special issue provide numerous examples. Readers of this journal are already familiar with the foundations and elements of behavior-based safety (BBS; also known as behavioral safety) as one effective approach. From its earliest inception and demonstrations (e.g., Komaki, Barwick, & Scott, 1978), BBS was intended to be an application of basic behavioral principles to positively affect not only worker behavior but also environmental conditions, system failures, and leadership decisions (McSween & Moran, 2017).

Unfortunately, the growth and evolution of BBS has resulted in some unrecognizable forms of BBS—some with unscrupulous practices—that have been meet with skepticism and criticism (e.g., Howe, 2001) in part because these approaches have focused too much on personal safety surrounding individual workers. For example, incentive systems, featuring great prizes and celebrations, were heralded as effective and efficient ways to promote safe behavior and discourage unsafe and at-risk behavior of workers. However, these practices were often implemented without regard to other organizational factors or other effective methods to eliminate or mitigate hazards and risks. Furthermore, by establishing rewards for achieving safety outcomes (i.e., fewer injuries), some ill-conceived incentive schemes function only to reinforce undesirable consequences, such as underreporting of injuries and, consequently, violating whistleblower and recordkeeping rules (OSHA, 2012). Although behavioral safety tends to focus on monitoring and reinforcing safe behavior of workers to achieve improved safety outcomes, long-term success requires comprehensive understanding and accounting of the roles of all other individuals in the organization, including coworkers, supervisors, managers, and executives.

As Ludwig (2017-a) explains, it is true that there are important differences between the contexts of personal safety and process safety; however, this does not mean necessarily that a fundamentally different foundation or approach is needed for PSM. Behavior analysis has already provided the intervention technologies necessary to make important contributions to PSM. For example, effective tools used for behavioral risk identification, such as PIC/NIC analysis (Agnew & Daniels, 2010) and the Performance Diagnostic Checklist (Austin, 2000; see also Martinez-Onstott, Wilder, & Sigurdsson, 2016, for a safety version) can be easily adapted to help safety professionals pinpoint critical process safety behaviors, including those behaviors and decisions of supervisors, managers, and executives. Rodriguez, Bell, Brown, and Carter (2017) and Hyten and Ludwig (2017) also show how behavioral training methods, such as fluency building, can be used to enhance skills and competence to reduce human error in performance and decision making.

As many of the articles in this special issue document, these behavioral methods are not entirely novel, and they have been known to be effective in many different organizational settings. Furthermore, as illustrated by Rodriguez et al. (2017) behavioral systems analysis and BBS approaches are synergistic with more traditional safety approaches (e.g., human factors) to reduce human error. An important common theme across several articles in this special issue is that these behavioral solutions are potentially useful not only for addressing workers' safety-related performance, but also for addressing the various roles of supervisors and other leaders within an organization.

McSween and Moran (2017) further describe how the components of a well-designed BBS program, which includes the components of pinpointing, observation, analysis, feedback, and reinforcement, already provides the tools and methods to address process failures with the same systematic rigor. Likewise, in their analysis of the behavioral contingencies that govern leadership safety-related decisions and actions, Gravina and Austin (2017) illustrated other examples of how a behavioral science approach to safety, including the tools and methods of BBS, can be expanded to address leader performance. Another good example is Zohar's (2002) study, which demonstrated the applicability and effectiveness of BBS practices to increase the frequency of supervisors' safety-related communications in what has been called "leader-based safety" (Kines et al., 2010).

Another promising intervention approach that is novel (at least among mainstream safety training practices) is teaching the principles of behavioral science to organizational leaders. Gravina et al. (2017) describe the benefits of a well-educated and coached leader through a "consultant workshop model," which gives leaders the knowledge and skills needed to create the behavioral supports necessary to sustain an effective PSM program. This approach goes beyond more traditional and common safety training approaches by focusing on analyzing and managing the behavioral aspects of leadership performance, which is considered throughout this special issue as being the main driver of process safety.

A call for research

Despite the strong scientific foundations of behavior analysis and proven successes in improving the human condition in many different arenas, greater efforts are needed to conduct more research and disseminate the findings to the wider occupational safety community. Lebbon and Sigurdsson (2017) identify several other important research opportunities for those working in the field of behavior analysis to address some process safety challenges. In my opinion, one of the most promising areas for impact is behavioral economics. Lebbon and Sigurdsson make a strong case for the relevance of delay and probability discounting and other behavioral economic theories of decision making in understanding factors that influence safety behavior in occupational settings, and they specify several avenues for continued research. Reward discounting by the consequence dimensions of delay and probability has been studied extensively in behavior analysis, but unfortunately these contributions, despite their relevance to occupational safety (see Reynolds & Schiffbauer, 2004), have not yet made their way into the mainstream safety literature.

Finally, Wirth and Sigurdsson (2008) provided a comprehensive research agenda surrounding many topics germane to the elements and practices of behavioral safety. For example, sample research questions were proposed in various categories including risk analysis and pinpointing, goal setting, training and prompting, observation and measurement, feedback, and rewards and incentives. Additional research topics were provided for methodological issues, safety culture and climate, and the integration of behavioral safety with other safety initiatives (e.g., PSM). Because appreciable overlap exists between behavioral safety and process safety, as demonstrated throughout this special issue, the research topics detailed by Wirth and Sigurdsson remain timely and relevant to PSM.

Final thoughts

The collection of articles in this special issue on process safety illustrates the many ways in which the field of behavior analysis offers a set of scientifically tested principles and technologies that can provide a solid framework for guiding research, practice, and policymaking in occupational safety and health. I have attempted to highlight a few specific topics raised by the contributors that I believe can be addressed uniquely and expertly by behavior analysis; however, success will depend on empirically validating the concepts, principles, and methods. Behavioral safety experts and consultants will play a vital role in this success. As evidence of their demonstrated successes over time, these professionals have established many positive and lasting relationships with workers, other safety professionals, and business leaders across many different industry sectors. The trust and cooperation gained can be leveraged into expanded partnerships with behavioral researchers in academia and government to conduct well-designed and well-executed applied research studies. Furthermore, these partnerships will allow access to the extensive data collected by companies on safety outcomes and other metrics associated with personnel, organizational, and environmental factors. These largely untapped data sources may prove to be a gold mine of important empirical evidence of tried and tested intervention approaches and, as Wagner (2014) pondered, suitable for predictive analytics and related approaches help predict when injuries or disasters are eminent.

One note of caution: Although the articles in this special issue present a wealth of information on the applicability of behavior analysis principles and methods to process safety, much of it may be inaccessible to the wider audience of professional and academic safety experts, not to mention workers and their employers. The jargon of behavioral science is highly technical, and many of the complex behavioral concepts and processes might not be readily understood by individuals without further education or training in behavior analysis. It behooves our field to consider ways to repackage this material and disseminate it to a broader audience of industrial hygienists, safety engineers, business leaders, and other occupational safety and public health experts to better promote the use of behavioral science and technologies in process safety programs.

On the other hand, this special issue will serve as a nice introduction to the topic of process safety for individuals with a strong background in behavioral sciences. The collection of articles demonstrates that the field of behavior analysis, well-established with scientifically validated behavioral principles and intervention technologies, can help bridge the gap between theory and practice on matters related to process safety. The successes of behavior analysis in improving the human condition in many different arenas suggest similar promises for improving the safety and well-being of workers.

Author note

The findings and conclusions in this report are those of the author and do not necessarily represent the views of the National Institute for Occupational Safety and Health.

References

Agnew, J., & Daniels, A. (2010). *Safety by accident: Leadership practices that build a sustainable safety culture*. Atlanta, GA: Performance Management Publications.

American Institute of Chemical Engineers (AIChE). (2016). *Guidelines for implementing process safety management* (2nd ed.). New York, NY: John Wiley and Sons.

Austin, J. (2000). Performance analysis and performance diagnostics. In J. Austin & J. E. Carr (Eds.), *Handbook of applied behavior analysis* (pp. 321–349). Reno, NV: Context Press.

Chemical Safety Board (2007). *Investigation report: Refinery explosion and fire* (No. 2005-04-I-TX). Author. Retrieved from http://www.csb.gov/assets/1/19/CSBFinalReportBP.pdf

Chemical Safety Board (2016). *Investigation report executive summary: Drilling rig explosion and fire at the Macondo Well* (No. 2010-1-I-OS. U.S). Author. Retrieved from http://www.csb.gov/macondo-blowout-and-explosion/

Glendon, I. (2008). Safety culture: A snapshot of a developing concept. *Journal of Occupational Health and Safety*, *24*(3), 179–189.

Glenn, S. S. (1988). Contingencies and metacontingencies: Toward a synthesis of behaviour analysis and cultural materialism. *The Behavior Analyst, 11*, 161–179. doi:10.1007/BF03392470

Glenn, S. S., & Malott, M. E. (2004). Complexity and selection: Implications for organizational change. *Behavior and Social Issues, 13*, 89–106. doi:10.5210/bsi.v13i2.378

Gravina, N., Cummins, B., & Austin, J. (2017). Leadership's role in process safety: An understanding of behavioral science is needed. *Journal of Organizational Behavior Management, 37*(3–4), 316–331.

Heinrich, H. W. (1931). *Industrial accident prevention: A scientific approach*. New York, NY: McGraw-Hill.

Hofmann, D., Burke, M., Zohar, D. (2017). 100 years of occupational safety research: From basic protections and work analysis to a multilevel view of workplace safety and risk. *Journal of Applied Psychology, 102*(3), 375–388.

Howe, J. (2001). *Warning! Behavior-based safety can be hazardous to your health and safety program!: A union critique of behavior-based safety*. UAW, International Union. Retrieved from http://www.uawlocal974.org/BSSafety/Warning!_Behavior-Based_Safety_Can_Be_Hazardous_To_Your_Health_and_Safety_Program!.pdf

Hyten, C., & Ludwig, T. (2017). Complacency in process safety: A behavior analysis toward prevention strategies. *Journal of Organizational Behavior Management, 37*(3–4) 240–260.

International Atomic Energy Agency. (1992). IAEA Report INSAG-7 Chernobyl Accident: Updating of INSAG-1 Safety Series (No.75-INSAG-7). Vienna, Austria: Author.

Kines, P., Andersen, L. P. S., Spangenberg, S., Mikkelsen, K. L., Dyreborg, J., & Zohar, D. (2010). Improving construction site safety through leader-based verbal safety communication. *Journal of Safety Research, 41*, 399–406. doi:10.1016/j.jsr.2010.06.005

Komaki, J., Barwick, K. D., & Scott, L. R. (1978). A behavioral approach to occupational safety: Pinpointing and reinforcing safety performance in a food manufacturing plant. *Journal of Applied Behavior Analysis, 63*, 434–445.

Kumar, S. (2014). Evolution of process safety management. *Indian Chemical Engineer, 56*(1), 61–70.

Lebbon, A., & Sigurdsson, S. (2017). Behavioral perspectives on variability in human behavior as part of process safety. *Journal of Organizational Behavior Management, 37*(3–4), 261–282.

Ludwig, T. (2017-a). EDITORIAL: Process safety: Another opportunity to translate behavior analysis into evidence-based practices of grave societal value. *Journal of Organizational Behavior Management, 37*(3–4), 221–223.

Ludwig, T. (2017-b). Process safety behavioral systems: Behaviors interlock in complex process safety meta-contingencies. *Journal of Organizational Behavior Management, 37*(3–4), 224–239.

Manuele, F. A. (2014). *Heinrich revisited: Truisms or myths* (2nd ed.). Itasca, IL: National Safety Council.

Martinez-Onstott, B., Wilder, D., & Sigurdsson, S. (2016). Identifying the variables contributing to at-risk performance: Initial evaluation of the Performance Diagnostic Checklist–Safety (PDCSafety). *Journal of Organizational Behavior Management, 36*, 80–93. doi:10.1080/01608061.2016.1152209

Mathis, T. L. (2013, October 14). Is the term "accident still acceptable? *EHS Today*. Retrieved from http://ehstoday.com/safety/term-accident-still-acceptable

McSween, T., & Moran, D. J. (2017). Assessing and Preventing Serious Incidents with Behavioral Science: Enhancing Heinrich's Triangle for the 21st Century. *Journal of Organizational Behavior Management, 37*(3–4), 283–300.

Occupational Safety and Health Administration (1994). *Process Safety Management Guidelines for Compliance, 3133 (reprinted)*. Retrieved from https://www.osha.gov/Publications/osha3133.html

Occupational Safety and Health Administration (2000). *Process Safety Management, 3132 (reprinted).* Retrieved from https://www.osha.gov/Publications/osha3132.html

Occupational Safety and Health Administration (2012, March 12). *Employer safety incentive and disincentive policies and practices.* Memorandum. Retrieved from https://www.osha.gov/as/opa/whistleblowermemo.html

Reynolds, B., & Schiffbauer, R. M. (2004). Impulsive choice and workplace safety: A new area of inquiry for research in occupational settings. *The Behavior Analyst, 27,* 239–246. doi:10.1007/BF03393183

Rodriguez, M., Bell, J., Brown, M., & Carter, D. (2017). Integrating behavioral science with human factors to address process safety. *Journal of Organizational Behavior Management, 37*(3–4), 301–315.

Schneider, B., Gonzalez Roma, V., Ostroff, C., West, M. (2017). Organizational climate and culture: Reflections on the history of the constructs in the Journal of Applied Psychology. *Journal of Applied Psychology, 102*(3), 468–482.

Skinner, B. F. (1953). *Science and human behavior.* New York, NY: Macmillan Company.

Skinner, B.F. (1981). Selection by consequences. *Science, 213*(4507), 501–504.

Wagner, G. R. (2014, October 2). *Can predictive analytics help reduce workplace risk?* NIOSH Science Blog. Retrieved from https://blogs.cdc.gov/niosh-science-blog/2014/10/02/pa/

Wirth, O., & Sigurdsson, S. (2008). When workplace safety depends on behavior change: Topics for behavioral safety research. *Journal of Safety Research, 39,* 589–598. doi:10.1016/j.jsr.2008.10.005

Zohar, D. (2002). Modifying supervisory practices to improve sub-unit safety: A leadership-based intervention model. *Journal of Applied Psychology, 87,* 156–163. doi:10.1037/0021-9010.87.1.156

Index